Nelson

MATH*FOCUS* 4

Senior Author and Senior Consultant
Marian Small

Authors
Jack Hope
Wendy Klassen
Marian Small
Rosita Tseng Tam
Stella Tossell

Assessment Consultant
Sandra Carl Townsend

THOMSON

NELSON

Australia   Canada   Mexico   Singapore   Spain   United Kingdom   United States

THOMSON

———★———

NELSON

Nelson Math Focus 4

Senior Author and Senior Consultant
Marian Small

Authors
Jack Hope, Wendy Klassen, Marian Small, Rosita Tseng Tam, Stella Tossell

Contributing Author
Kathleen Kacuiba

Assessment Consultant
Sandra Carl Townsend

Director of Publishing
Beverley Buxton

General Manager, Mathematics, Science, and Technology
Lenore Brooks

Publisher, Mathematics
Colin Garnham

Associate Publisher, Mathematics
Sandra McTavish

Managing Editor, Development
David Spiegel

Product Manager
Linda Krepinsky

Program Manager
Mary Reeve

Developmental Editors
Colin Bisset
First Folio Resource Group, Inc.:
  Brenda McLoughlin
  Bradley Smith
  Robert Templeton
  Susan Woollam

Assistant Editors
Linda Watson, First Folio Resource Group, Inc.
Carmen Yu

Editorial Assistant
Caroline Winter

Executive Director, Content and Media Production
Renate McCloy

Director, Content and Media Production
Linh Vu

Content Production Editor
Carolyn Pisani

Copy Editor
Susan McNish

Proofreaders
David Gargaro
Sabrina Mansour
Natalie Russell
Caroline Winter

Indexer
Noeline Bridge

Production Manager
Cathy Deak

Senior Production Coordinator
Sharon Latta Paterson

Design Director
Ken Phipps

Interior Design
Kyle Gell Design

Cover Design
Will Bache

Cover Image
Stephen Frink/Getty Images

Illustrators
Deborah Crowle
Kyle Gell
Kathy Karakasidis
Heather Moon
Dave Whamond

Compositor
Kyle Gell

Photo/Permissions Researcher
Alison Lloyd

Printer
RR Donnelley/Willard

COPYRIGHT © 2008 by Nelson, a division of Thomson Canada Limited.

ISBN-13: 978-0-17-632449-0
ISBN-10: 0-17-632449-6

Printed in the United States.
1 2 3 4 11 10 09 08

For more information contact Thomson Nelson, 1120 Birchmount Road, Toronto, Ontario, M1K 5G4. Or you can visit our Internet site at http://www.nelson.com

Every effort has been made to trace ownership of all copyrighted material and to secure permission from copyright holders. In the event of any question arising as to the use of any material, we will be pleased to make the necessary corrections in future printings.

Advisory Panel

The authors and publisher gratefully acknowledge the contributions of the following educators.

James Beres
Teacher
Stirling School
Westwind School Division #74
Stirling, Alberta

Bob Boyechko
Teacher
St. Elizabeth Seton School
Edmonton Catholic Schools
Edmonton, Alberta

Carol Brydon
Teacher and former Mathematics
Consultant
Calgary Catholic School Board
Calgary, Alberta

Andrea Clarke
Teacher
Earl Grey Elementary
Calgary Board of Education
Calgary, Alberta

Deb Colvin-MacDormand
Teacher
Jackson Heights School
Edmonton Public Schools
Edmonton, Alberta

Edna Dach
Director
Elk Island Public Schools
Sherwood Park, Alberta

Monica Devereux
Teacher
Alex Munro Elementary
Calgary Board of Education
Calgary, Alberta

Shona Dobrowolski
Mathematics Consultant
Okotoks, Alberta

Lenée Fyfe
Teacher
Park Meadows School
Lethbridge School Division No. 51
Lethbridge, Alberta

Peggy Gerrard
Vice Principal
Percy Pegler School
Foothills School Division
Okotoks, Alberta

Jennifer Gluwchynski
Teacher
St. Elizabeth Seton School
Edmonton Catholic Schools
Edmonton, Alberta

Elizabeth Grill-Donovan
Vice Principal
St. Cyril School
Calgary Roman Catholic Separate
School District #1
Calgary, Alberta

Kathy (Canchun) Gu
Teacher
Caernarvon Elementary School
Edmonton Public School Board
Edmonton, Alberta

Laurie Kardynal-Bahri
Consultant
Elk Island Catholic School
Division #41
Sherwood Park, Alberta

Carla Kozak
Consultant
Edmonton Public Schools

Rebecca Kozol
Teacher
Yennadon Elementary
SD #42
Maple Ridge, British Columbia

Jacinthe Lavoie
Enseignante
École Elizabeth Rummel School
Canadian Rockies Public Schools
Canmore, Alberta

Hugh MacDonald
Principal
St. Elizabeth Seton School
Edmonton Catholic Schools
Edmonton, Alberta

Bernard MacGregor
Assistant Principal
St. Maria Goretti Elementary
School
Edmonton Catholic Schools
Edmonton, Alberta

Moyra Martin
Principal
Cardinal Newman
Elementary/Junior High School
Calgary Catholic School District
Calgary, Alberta

Nicole Merz
AISI Consultant
Calgary Board of Education
Calgary, Alberta

Mary Anne Nissen
Consultant
Edmonton, Alberta

Charlotte Oliver
Grade 5 Teacher
Glamorgan Elementary School
Calgary Board of Education
Calgary, Alberta

Janet Reid
Elementary Teacher
Alex Munro Elementary
Calgary Board of Education
Calgary, Alberta

Doug Super
Mathematics Teacher
Mulgrave School
West Vancouver, British
Columbia

Bryan Szumlas
Principal
Our Lady of Fatima
Calgary Roman Catholic Separate
School District #1
Calgary, Alberta

Gerry Varty
AISI/Math Coordinator
Wolf Creek Public Schools
Ponoka, Alberta

Paula Watson
Grade 3 Teacher
Monsignor J.J. O'Brien School
Calgary Catholic School Board
Calgary, Alberta

Tammy Watz
Teacher
Ben Calf Robe/St. Clare School
Edmonton Catholic Schools
Edmonton, Alberta

Tracy Welke
Mathematics Teacher
Vegreville, Alberta

Aboriginal Consultants

Nicole Bell
Assistant Professor
Trent University
Peterborough, Ontario

Fred N. Butler
Principal
Sakku School
Kivalliq School Operations
Coral Harbour, Nunavut

Jennifer Hingley
Project Leader, First Nation and
Métis Content and Perspectives
Okiciyapi Partnership
Saskatoon Public Schools and
Saskatoon Tribal Council

Jacqui Lavalley
Ojibwe Tradition/Cultural
Instructor
First Nations School of Toronto
Toronto District School Board
Toronto, Ontario

Susie Robinson
Cree Language Consultant
Aboriginal Learning Services
Edmonton Catholic Schools
Edmonton, Alberta

Michael D. Thrasher-Kawhywaweet
Elder
Victoria, British Columbia

Equity Reviewer

Mary Schoones
Educational Consultant/
Retired Teacher
Ottawa-Carleton District School
Board
Ottawa, Ontario

Literacy Consultants

Vicki McCarthy
Language and Literacy
Consultant
Vancouver School District
Vancouver, British Columbia

Melanie Quintana
Teacher
St. Julia Catholic Elementary
School
Dufferin-Peel Catholic District
School Board
Mississauga, Ontario

Contents

Chapter 3 Addition and Subtraction   64

Chapter 4 Data Relationships   102

Chapter 5

2-D Geometry   140

Chapter 6

Multiplication and Division Facts   166

Chapter 7   Fractions and Decimals   204

Chapter 8

Measurement   258

Chapter 11

3-D Geometry   376

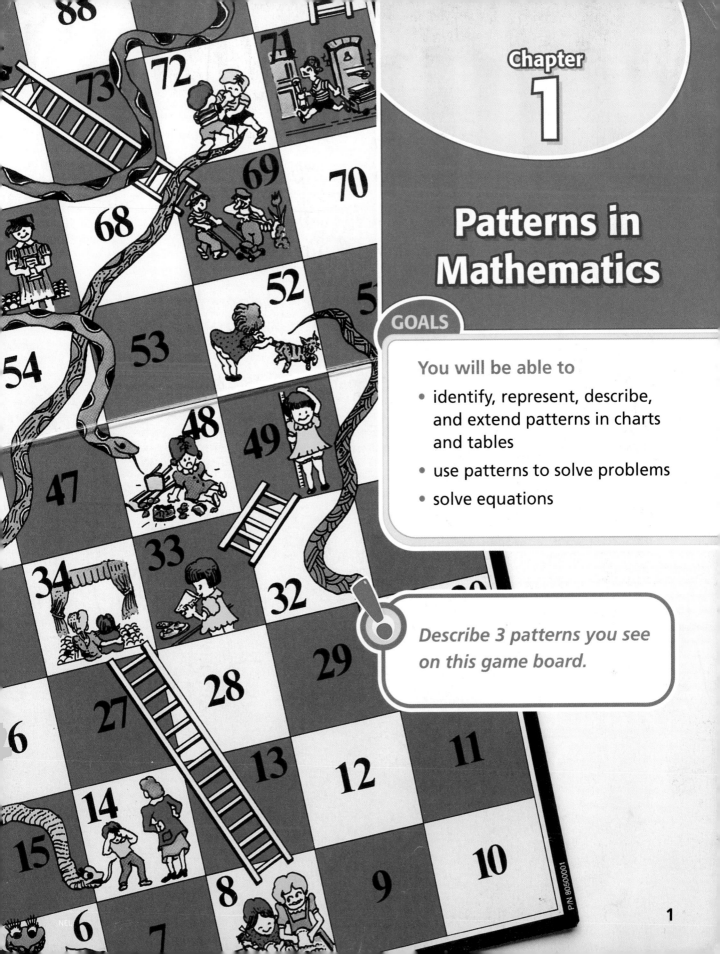

Patterns in Mathematics

GOALS

You will be able to

- identify, represent, describe, and extend patterns in charts and tables
- use patterns to solve problems
- solve equations

Describe 3 patterns you see on this game board.

Getting Started

You will need
- counters

Reading Strategy

Flip through the Student Book. What special features can you find?

Comparing Patterns

Cole has 30 marbles and Kate has 12.
Cole gives Kate 3 of his marbles at a time.

? **Will Cole and Kate ever have the same number of marbles?**

Cole counts his marbles: 30, 27, 24, …
Kate counts her marbles: 12, 15, 18, …

A. Model Cole's and Kate's actions using counters.

B. Describe each pattern.

C. How are the 2 patterns alike? How are they different?

D. Extend each pattern for 3 more numbers.

E. Will Cole and Kate ever have exactly the same number of marbles? Explain.

F. Suppose Cole gives Kate 4 marbles at a time instead of 3. Repeat Parts A to E.

What Do You Think?

Do you *agree* or *disagree* with each statement? Explain your thinking.

1. There are different kinds of patterns.

2. A pattern that starts with the numbers 10, 20, ... can be continued in only one way.

3. A 100 chart shows only increasing patterns.

1	2	3	4	5	6	7	8	9	10
11	12	13	14	15	16	17	18	19	20
21	22	23	24	25	26	27	28	29	30
31	32	33	34	35	36	37	38	39	40
41	42	43	44	45	46	47	48	49	50
51	52	53	54	55	56	57	58	59	60
61	62	63	64	65	66	67	68	69	70
71	72	73	74	75	76	77	78	79	80
81	82	83	84	85	86	87	88	89	90
91	92	93	94	95	96	97	98	99	100

4. There is more than one way to solve the equation $4 + \blacksquare = 11$.

Patterns in an Addition Table

You will need
- Addition Tables (blackline master)
- pencil crayons

> **GOAL**
>
> Identify, describe, and complete patterns in an addition table.

? **What number patterns can you use to complete the addition table?**

column

+	0	1	2	3	4	5	6	7	8	9
0	0	1	2	3	4	5	6	7	8	9
1		2	3	4	5	6	7	8	9	10
2		3	4		6	7		9		11
3	3	4	5	6	7	8	9	10	11	
4		5		7	8	9	10	11	12	13
5	5	6	7		9	10	11	12	13	14
6	6	7		9	10		12	13	14	
7	7	8	9	10		12	13	14	15	16
8		9	10		12		14	15	16	17
9	9	10	11	12	13		15	16	17	18

← row

↖ diagonal

pattern rule

A description of how a pattern starts and how it continues.

For example, for the pattern 24, 27, 30, 33, ..., here is the pattern rule: Start with 24 and add 3 each time.

Tien's Pattern

I found an increasing pattern I can use. In the yellow column, each number is 1 more than the number above it.

The number pattern is 5, 6, 7, 8, 9, 10, ...

Here is the **pattern rule**:

Start with 5 and add 1 each time.

Lang's Pattern

I found a decreasing pattern. Starting with 16 and going up the green diagonal, each number is 2 less. The number pattern is 16, 14, 12, 10, ...
This is the pattern rule:
Start with 16 and subtract 2 each time.

A. Complete the yellow column using Tien's pattern rule.

B. Complete the green diagonal using Lang's pattern rule.

C. Choose a row with missing numbers. Describe the pattern going from right to left. Is it a decreasing pattern or an increasing pattern?

D. Choose a column with missing numbers. Describe the pattern going from top to bottom. Is it a decreasing pattern or an increasing pattern?

E. Describe 2 other patterns in the addition table.

F. Complete the addition table using patterns.

Reflecting

G. Look at the patterns in each row going from left to right. How are these patterns alike? How are they different?

H. Which directions do the decreasing patterns go on the addition table?

I. How did you use patterns to complete the addition table?

Checking

1. Complete an addition table like this using patterns. Use at least one pattern from a row, a column, and a diagonal. Describe the patterns you used.

+	2	4	6	8	10	12	14
2	4	6	8	10	12		
4	6		10		14	16	
6	8	10	12		16	18	
8		12		16		20	22
10			16		20	22	
12	14		18		22	24	26
14	16	18				26	

Practising

2. Complete an addition table like this using patterns. Use at least one pattern from a row, a column, and a diagonal. Describe the patterns you used.

+	2	4	6	8	10	12	14
1	3	5	7	9		13	
3	5	7		11	13	15	17
5	7	9	11	13	15		19
7		11	13		17	19	21
9			15		19	21	
11	13			19	21		25
13		17				25	

3. Complete an addition table like this using patterns. Describe the patterns you used.

+	10	20	30	40	50	60	70
1	11		31	41	51		71
2	12	22	32			52	62
3	13		33		53	63	73
4	14	24		44		64	
5			35		55	65	
6	16			46	56		76
7		27	37		57	67	

4. Lindsay says the pattern in each row of an addition table can be described as an increasing and as a decreasing pattern. Do you agree or disagree? Explain with examples.

Patterns in Charts

Number of players: 2 or more

How to play: Create number patterns from pattern rules and cover numbers on a 100 chart to show the patterns.

- **Step 1** Shuffle the number cards and place them face down in a pile. Give one 100 chart to each player.

- **Step 2** On your turn, take 2 cards. Use the numbers to make a pattern rule.

- **Step 3** Place coloured counters on your 100 chart to cover all of the numbers in your pattern.

- **Step 4** Each player takes 2 turns. On each turn, use a different-coloured counter on your 100 chart so you can see the new pattern.

The winner is the player with the greatest number of counters on his or her 100 chart. Count only the top counter on any number.

Julia's Turn

Here is my pattern rule: Start with 6 and add 7 each time.

6	7

1	2	3	4	5	◯	7	8	9	10
11	12	◯	14	15	16	17	18	19	◯
21	22	23	24	25	26	◯	28	29	30
31	32	33	◯	35	36	37	38	39	40
◯	42	43	44	45	46	47	◯	49	50
51	52	53	54	◯	56	57	58	59	60
61	◯	63	64	65	66	67	68	◯	70
71	72	73	74	75	◯	77	78	79	80
81	82	◯	84	85	86	87	88	89	◯
91	92	93	94	95	96	◯	98	99	100

Extending Patterns in Tables

You will need
- linking cubes
- square tiles
- toothpicks

Inuksuk (plural: Inuksuit)

GOAL

Use tables to identify and extend patterns.

An inuksuk is a marker or signpost made of rocks. Kate collected 15 small rocks and 26 large rocks.

? **Does Kate have enough small rocks to make 6 inuksuit like this one?**

Kate's Inuksuk Project

Understand
I need to find out how many small rocks I need for making 6 inuksuit.

Make a Plan
I'll make models of the inuksuk using linking cubes.

large rock   small rock

I'll use a table to solve the problem.

Carry Out the Plan

After making 3 inuksuit, I notice a pattern in the second column: 3, 6, 9, …. I'll extend the pattern to complete the table.

Look Back

For 6 inuksuit, I need 18 small rocks. I have only 15 small rocks, so I don't have enough.

Small Inuksuk Rocks

Number of inuksuit	Total number of small rocks
1	3
2	6
3	9
4	12
5	15
6	18

Reflecting

A. Describe the pattern in the second column of Kate's table.

B. How did the table help Kate solve the problem?

Checking

1. a) Kate has 26 large rocks. Does she have enough large rocks to make 6 inuksuit? Use a table.

Large Inuksuk Rocks

Number of inuksuit	Total number of large rocks
1	
2	
3	

b) Describe the pattern in the second column.

Practising

2. There is an inuksuk on the Nunavut flag.

   a) How many rocks are needed to make 7 of these inuksuit? Use a table.

   Inuksuk Rocks

Number of inuksuit	Total number of rocks
1	
2	
3	

   b) Describe the pattern in the second column of the table.

3. Joshua made this shape pattern.

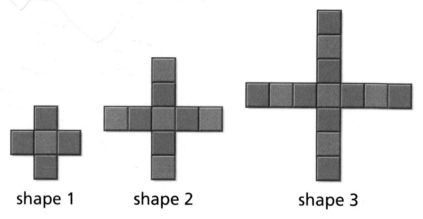

shape 1   shape 2   shape 3

   a) If Joshua continues this pattern, how many squares will be in shape 6? Use a table.

   Shape Squares

Shape number	Number of squares in the shape
1	5
2	9
3	

   b) Describe the pattern in the second column.

10

4. Jay made this pattern out of toothpicks.

Toothpick Pattern	
Number of shapes	Total number of toothpicks
1	3
2	7
3	10
4	

a) How many toothpicks did he use for all 7 shapes?
b) Describe the pattern in the second column of the table.
c) Jay has 35 toothpicks. If he continues his pattern until all the toothpicks are used, how many shapes will he have? Explain your thinking.

5. Cara is making this shape pattern using square tiles. If she continues this pattern until the last shape has 1 square, how many shapes will the pattern have?

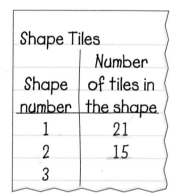

Shape Tiles	
Shape number	Number of tiles in the shape
1	21
2	15
3	

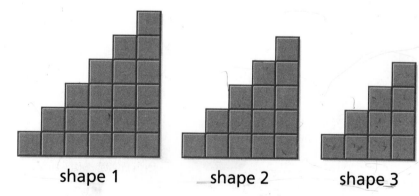

shape 1   shape 2   shape 3

6. a) Build a model of an inuksuk using linking cubes,
b) How many of each size of rock would you need to make 6 of your inuksuit?
c) Do you need to model every shape to answer part (b)? Explain.

Representing Patterns

You will need
- craft sticks
- base ten blocks

GOAL

Use models to represent and extend patterns.

Joshua is making square picture frames using craft sticks.
He has 30 craft sticks. The table shows the pattern for the number of craft sticks needed to make the frames.

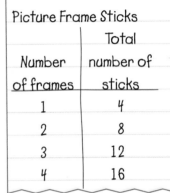

Picture Frame Sticks	
Number of frames	Total number of sticks
1	4
2	8
3	12
4	16

? How many frames can Joshua make using 30 sticks?

Joshua's Pattern

I'll model the pattern in the table using sticks.
I'll add 4 more sticks each time.
I'll extend the model of the pattern to see how many frames I can make using 30 sticks.

I can make 7 frames and I'll have 2 sticks left over.

Reflecting

A. How do Joshua's model and the table show the same information?

B. How did Joshua use the model to solve the problem?

Checking

1. Ayana made a different pattern with frames.

a) Model the pattern in the table.

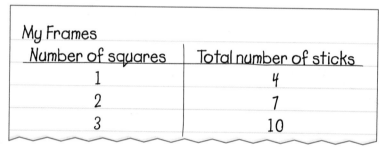

My Frames	
Number of squares	Total number of sticks
1	4
2	7
3	10

b) Describe how the model changes.

c) How many squares long can Ayana make the frame if she has 30 craft sticks?

Practising

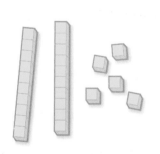

2. Mario's class started with 19 chairs in a game called Musical Chairs. Mario took away 3 chairs each time the music stopped.

a) Model the pattern in the table.

Musical Chairs	
Stop number	Number of chairs used
0	19
1	16
2	13

b) Describe how the model changes.

c) How many times did the music stop before there was only 1 chair left?

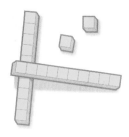

3. Karen made a training schedule to prepare for a kickboxing tournament.
   a) Model the pattern in the table.

Training Schedule				
Week number	1	2	3	4
Number of sit-ups each day	10	25	40	55

   b) Describe how the model changes.
   c) How many sit-ups will she be doing each day in 7 weeks?

4. Bree is making 5-sided frames. She wants to find out how many sticks she needs to make 7 frames. She uses a model and a table to find out. Shayla noticed errors in Bree's model and table.

My Frames	
Number of frames	Total number of sticks
1	5
2	10
3	15
4	21
5	26
6	31
7	36

   a) Describe the errors in Bree's model and table.
   b) Was it easier for you to see the error in the model or in the table? Explain.

5. a) Create your own pattern in a table with 5 rows.
   b) Model your pattern.
   c) How are your number pattern and model alike? How are they different?

Number Chains

1. This is a number chain. On each triangle, the numbers on the corners add up to the number inside.

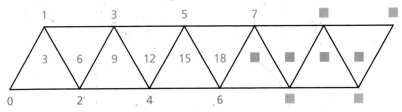

a) Describe the pattern of the red numbers.
What are the missing red numbers?

b) Describe the pattern of the blue numbers.
What are the missing blue numbers?

c) Describe the pattern of the green numbers.
What are the missing green numbers?

d) Describe the pattern in the zigzagging blue and green numbers.

e) For the last 4 triangles, calculate the sum of the 3 corners. What do you notice?

2. Repeat Question 1 for this number chain.

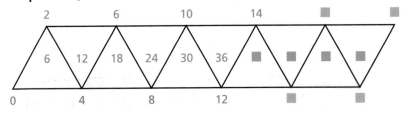

3. Repeat Question 1 for this number chain.

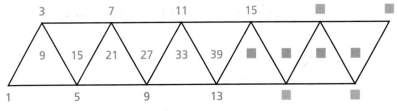

Mid-Chapter Review

Frequently Asked Questions

+	5	10	15	20	25	30
2	7	12	17	22	27	32
4	9		19	24	29	
6	11	16	21		31	
8		18		28	33	

Q: How can you identify, describe, and complete a pattern in a table?

A: Start with any row, column, or diagonal. Identify how the numbers in the pattern change. For example, in the 6+ row, the pattern is 6, 11, 16, 21 …. The numbers in the pattern increase by 5 each time.

To identify the missing number, add 5 to the number before: $21 + 5 = 26$.

To extend the pattern, add 5 to the last number: $31 + 5 = 36$.

Q: How can you use a model to represent and extend a pattern?

Number of frames	1	2	3	4		
Total number of sticks	3	6	9	12		

A: For example, how many sticks do you need to make 6 frames? Model the first 4 frames using craft sticks. Then extend the model. Count the number of frames and the total number of sticks at each step to extend the pattern in the table.

Number of frames	1	2	3	4	5	6
Total number of sticks	3	6	9	12	15	18

Practice

Lesson 1

1. Complete an addition table like this using patterns. Use at least one pattern from a row, a column, and a diagonal. Describe the patterns you used.

+	1	2	3	4	5	6	7
11	12		14	15	16		
22	23	24	25		27	28	
33	34		36		38	39	40
44		46		48		50	51

Lesson 2

2. Lana created this shape pattern from counters.

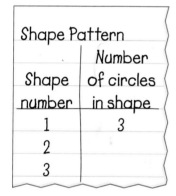

Shape Pattern

Shape number	Number of circles in shape
1	3
2	
3	

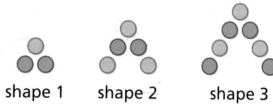

shape 1   shape 2   shape 3

a) If she continues this pattern, how many counters will be in shape 6? Use a table.

b) Describe the pattern in the second column.

Lesson 3

3. Shelby was playing a computer game. Every time she got a certain number of points, she moved up to the next level.

a) Model the pattern in the table using base ten blocks.

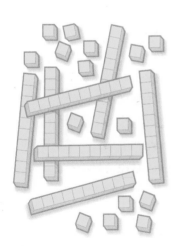

Computer Game Points					
Level	1	2	3	4	5
Total points	12	24	36	48	60

b) If Shelby's pattern continues, what will be her score on level 8?

Solving Problems Using Patterns

You will need
- a 100 chart
- Blank Calendar
  (blackline master)

GOAL

Use a pattern in a chart to solve a problem.

Here comes a parade of 100 clowns!
Every 2nd clown has a red nose.
Every 3rd clown wears glasses.

? How many clowns have a red nose and glasses?

Aneela's Solution

Understand
Some clowns have a red nose. Others have glasses. Some have both. How many have both?

Make a Plan
I need to count every 2nd and 3rd clown in a line of 100 clowns. I'll mark the numbers on a 100 chart.

Carry Out the Plan

\ is for red nose.

/ is for glasses.

After 3 rows, I see that every 6th number has both marks.
I'll circle every 6th number and count.

16 clowns have a red nose and glasses.

Reflecting

A. How did using a pattern make the problem easier to solve?

Checking

1. Every 5th clown in the parade of 100 clowns wears a hat. Every 2nd clown has a red nose. How many clowns have a hat *and* a red nose?

Practising

2. **a)** Every 3rd clown in the parade of 100 clowns wears glasses. Every 5th clown wears a hat. How many clowns have a hat *and* glasses?
   b) Every 2nd clown has a red nose. How many clowns have a hat, glasses, *and* a red nose?

MAY						
Sun.	Mon.	Tues.	Wed.	Thurs.	Fri.	Sat.
	1	2	3	4	5	6
7	8	9	10	11	12	13
14	15	16	17	18	19	20
21	22	23	24	25	26	27
28	29	30	31			

3. Shaun walks his dog every 3rd day. He goes for a run every 4th day. How many times in a month does Shaun walk his dog and go for a run on the same day?

4. Britney's family went skating together on a Tuesday. After that, Britney and her brother Liam skated every 2nd day. Their mother skated every 3rd day and their father skated every 4th day. On what day of the week did Britney's family next skate together?

January						
Sun.	Mon.	Tues.	Wed.	Thurs.	Fri.	Sat.
1	2	3	4	5	6	7
8	9	10	11	12	13	14
15	16	17	18	19	20	21
22	23	24	25	26	27	28
29	30	31				

5. Create and solve a problem that can be solved using a pattern in a chart or calendar.

Solving Equations

You will need
- number lines
- base ten blocks

Determine the missing number in an equation.

Kate's riding club is baking and freezing pies for a fundraiser. They started with 6 pies.
Then they baked the same number of pies each week.
The total number of pies at the end of each week made this pattern: 6, 11, 16, 21, 26, ….

equation

A mathematical sentence in which the value of the left side is the same as the value of the right side

 How many pies did Kate's riding club bake each week?

Tien's Solution

I'll write an **equation**. The missing number tells how much the pattern increases each time.

6 + ▊ = 11

I'll find 6 on a number line. Then I'll count the number of spaces to 11.

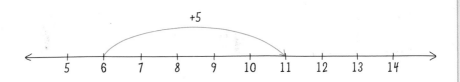

$$6 + 5 = 11$$

Kate's club baked 5 pies each week.

Lang's Solution

I'll write an equation and model it with base ten blocks.

$$21 + \blacksquare = 26$$

Left side   =   Right side

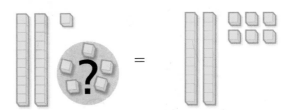

I'll add ones blocks until both sides have the same value.

I added 5 ones blocks.

So, $21 + 5 = 26$.

Kate's club baked 5 pies each week.

Reflecting

A. How did Tien use a number line to solve the problem?

B. What does $\blacksquare$ mean in Lang's equation?

Checking

1. Tara's family had 77 packages of frozen strawberries. Each week, they ate the same amount of strawberries. The number of packages in the freezer made this pattern.

   77, 73, 69, 65, 61 ...

   a) Write an equation with a missing number to represent how the pattern decreases.
   b) What is the missing number in your equation? Describe how you figured out the missing number.

Practising

2. Aaron started with a case of 52 apples. Each day he put one apple in each family member's lunch. The number of apples in the case made this pattern.

   52, 45, 38, 31, 24, ...

   a) How many people are in Aaron's family? Use an equation with a missing number to solve the problem.
   b) Describe how you figured out the missing number.

> **Reading Strategy**
>
> Read the question. Identify any words that you don't understand. Write a definition. Use a picture and an example.

3. The numbers in this pattern increase by different amounts each time.

   9, 11, 14, 18, 23, ...

   The missing numbers in the equations show how the pattern increases. What are the missing numbers?
   a) $9 + \blacksquare = 11$
   b) $11 + \blacksquare = 14$
   c) $14 + \blacksquare = 18$
   d) $18 + \blacksquare = 23$

4. Rebecca put the same amount of money into a savings account each week. The total number of dollars in her account each week made this pattern.

   11, 26, 41, 56, 71, 86, ...

   a) Write an equation with missing numbers to represent how the pattern changes.
   b) How much money did Rebecca save each week?

5. What is the missing number in each equation? Use a number line to help you.

   a) ■ + 7 = 16   d) 6 + 3 = ■
   b) 21 − ■ = 5   e) 27 = 11 + ■
   c) ■ − 9 = 35   f) 36 + ■ = 52

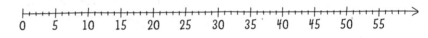

6. This pattern does not change by the same amount each time.

   59, 54, 52, 47, 45, 40, 38, ...

   a) Write equations with missing numbers to represent how the pattern changes.
   b) What are the missing numbers in your equations?
   c) Describe the pattern.

7. a) Describe what this equation means.
      ■ + 6 = 45
   b) What is the missing number in the equation? Describe how you figured it out.

Solving Problems with Equations

GOAL

Use equations to solve problems.

Cole is volunteering at a food co-op.
He is packaging almonds to make 450 g bags.
He has put 185 g in a bag.

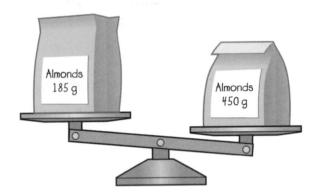

How many more grams of almonds does Cole need to add to the 185 g bag?

Cole's Solution

I'll use an equation to solve the problem.
■ is the number of grams of almonds I need
to add to the bag.

$185 + ■ = 450$

I'll use **guess and test** to figure out ■.

A. Solve the problem using Cole's strategy.

Reflecting

B. Why did Cole write 185 and ▢ on the same side of the equation?

C. How did writing an equation help Cole solve the problem?

Checking

1. Cole needs to make 250 g bags of raisins. So far he has 118 g in a bag.

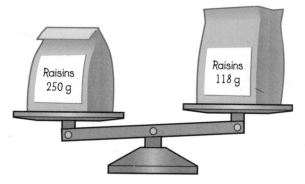

a) How many more grams of raisins does Cole need to add to the 118 g bag? Use an equation to solve the problem.

b) Describe how you figured out the missing number.

Practising

2. Tracy froze 35 kg of blueberries. Two months later, she had 7 kg left. How many kilograms of blueberries did she use? Use an equation to solve the problem.

3. Gabrielle and Hannah are delivering flyers. They both want to deliver the same number of flyers. How many flyers does Hannah need to add to her wagon?

4. Colin had 75 raffle tickets to sell. He has already sold some tickets, but he still needs to sell 36 more. How many raffle tickets has Colin sold already?

5. Simon wants to plant 175 corn plants. So far, he has planted 118. How many more corn plants does he need to plant?

6. Alissa beaded a necklace over 2 days. On the 1st day, she used 135 beads. When she was finished, she had used a total of 250 beads. How many beads did she use on the 2nd day?

7. Create and solve a problem for each number sentence.
   a) $92 + \blacksquare = 137$
   b) $78 - \blacksquare = 23$

Equations in a Story

GOAL

Create and solve equations to go with a story.

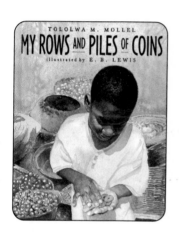

In *My Rows and Piles of Coins,* by Tololwa M. Mollel, a Tanzanian boy named Saruni saves his coins to buy a bicycle so that he can help his parents carry goods to market.

Saruni starts off with 5 ten-cent coins. When he is ready to buy the bicycle, he has 305 ten-cent coins. To find out how many coins Saruni saved, you can write and solve this equation: $5 + \blacksquare = 305$

? **What equations can you create and solve about saving money?**

Chapter Review

Frequently Asked Questions

Q: **How can you use an equation to solve a problem involving a missing number?**

A: First, write an equation for the problem. Then figure out the missing number in the equation.

For example, Malila is freezing 250 g bags of salal berries. She has 120 g in a bag. To figure out how many more berries she must add, write an equation. Use a symbol for the missing number.

$$120 + \blacksquare = 250$$

There are many ways to figure out the missing number.

- **Model** both sides of the equation using base ten blocks. Add base ten blocks until the value on each side is the same.

Left side   =   Right side

? =

- Use **guess and test** to figure out the missing number.
- Find 120 on a **number line**. Count the number of spaces to 250.

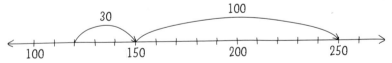

The missing number is 130. Malila needs to add 130 g of salal berries to the bag.

Practice

1. Complete an addition table like this using patterns.
   Describe the patterns you used.

+	2	4	6	8	10
10	12	14		18	20
11	13		17	19	
12		16	18		22
13			19		

Lesson 2

2. Olivia made this pattern using toothpicks.

shape 1   shape 2   shape 3   shape 4   shape 5   shape 6   shape 7

   a) How many toothpicks will she need for 10
      shapes? Use a table.
   b) Describe the pattern in the second column.

Toothpick Shapes

Total number of shapes	Total number of toothpicks
1	4
2	10
3	14

Lesson 3

3. Sean used a table to record how many people
   could sit at all the picnic tables in the park.

Picnic Tables

Number of picnic tables	Total number of people
1	8
2	16
3	24
4	32

   a) Model the pattern.
   b) If the pattern continues, how many people can
      sit at 7 picnic tables?

AUGUST						
Sun.	Mon.	Tues.	Wed.	Thurs.	Fri.	Sat.
		1	2	3	4	5
6	7	8	9	10	11	12
13	14	15	16	17	18	19
20	21	22	23	24	25	26
27	28	29	30	31		

Lesson 4

4. In August, Nicky did a crossword puzzle every 5 days. She did a word search puzzle every 3 days. How many times in August did she do both types of puzzles on the same day? _2_

Lesson 5

5. What is the missing number in each equation?
   a) $13 + 9 = 22$
   b) $27 = \blacksquare - 5$
   c) $45 + 15 = 60$
   d) $32 - \blacksquare = 25$

6. At the start of a game, Kelly gave each player the same amount of play money. The total number of dollars Kelly gave the players made this pattern:

   16, 32, 48, 64, ….

   a) Write an equation with a missing number to represent how the pattern changes.
   b) How much play money did each player get at the start of the game?

Lesson 6

7. Manuel is saving for a pair of running shoes that cost $89. He still needs $42. How much has he saved already? Use an equation to solve the problem. _47_

Lesson 7

8. Jade's goal is to save 500 pennies. She already has 256. How many more pennies does Jade need to meet her goal? Use an equation to solve the problem. _244_

What Do You Think Now?

Look back at **What Do You Think?** on page 3. How have your answers and explanations changed?

Chapter Task

✔ Did you include a diagram?

✔ Did you use tables to show number patterns?

✔ Did you describe the number patterns by writing pattern rules?

Number Patterns in Shape Patterns

Vasco created this shape pattern using pattern blocks.

shape 1   shape 2   shape 3   shape 4

 What number pattern can you see in the shape pattern?

Part 1

A. Make a model or sketch of shape 4.

B. Copy and complete a table to show the number pattern for Vasco's shape pattern.

C. Describe the pattern in the number of blocks.

D. How many blocks would be in shape 5 and shape 6?

Part 2

E. Create your own shape pattern using pattern blocks. Sketch and describe your shape pattern.

F. Describe a number pattern in your shape pattern.

G. How are your pattern and Vasco's pattern the same? How are the patterns different?

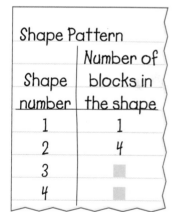

Shape Pattern

Shape number	Number of blocks in the shape
1	1
2	4
3	
4	

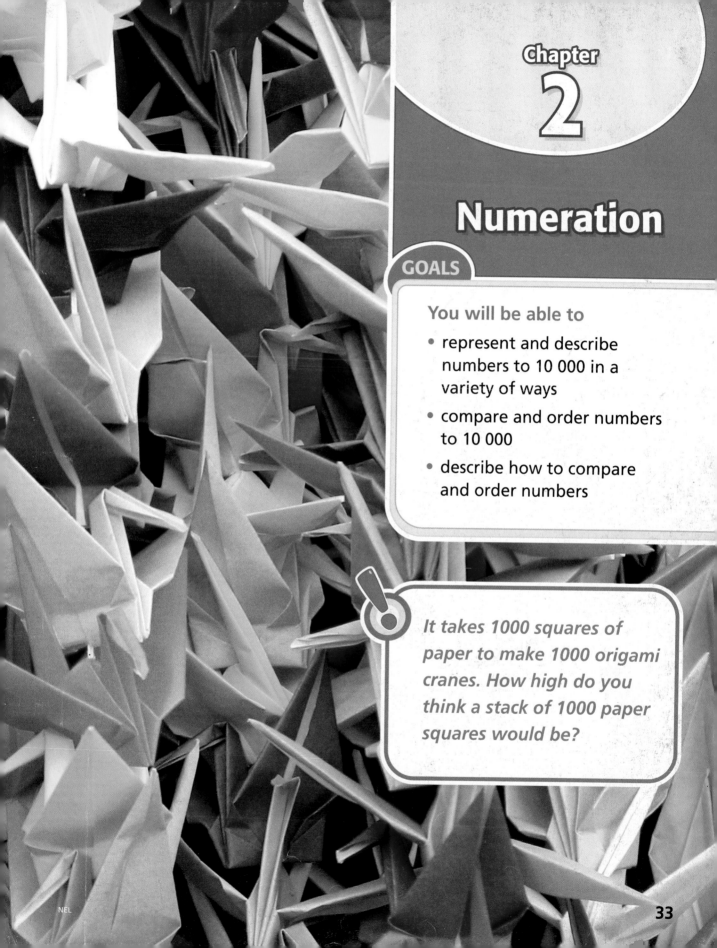

Numeration

GOALS

You will be able to

- represent and describe numbers to 10 000 in a variety of ways
- compare and order numbers to 10 000
- describe how to compare and order numbers

It takes 1000 squares of paper to make 1000 origami cranes. How high do you think a stack of 1000 paper squares would be?

Getting Started

Modelling Numbers

A man in India knows how to make the sounds of 326 mammals and birds.

? **How can you model 3-digit numbers like 326?**

Olivia's Model of 326

I modelled 326 using 11 base ten blocks.

Hundreds	Tens	Ones
▦ ▦ ▦	▯ ▯	▫ ▫ ▫ ▫ ▫ ▫

A. How does Olivia's model show that 326 is between 300 and 400?

B. Olivia's model for 326 uses 11 blocks. Model 326 using more than 11 blocks.

C. Use 11 blocks to model a 3-digit number you choose. Then model your number with more blocks.

D. Repeat Part C twice. Model a different number each time.

E. Which of the 3 numbers that you modelled is the greatest? How do you know?

F. Which number is the least? How do you know?

G. Which number is closest to 300? How do you know?

What Do You Think?

Do you *agree* or *disagree* with each statement?
Explain your thinking.

1. If one number is greater than another, you need more base ten blocks to model it.

2. The number ■84 can be greater than ■94 if you put the right numbers in the boxes.

3. If you can model a 3-digit number with 4 base ten blocks, you can also model it with 13 blocks.

Modelling Thousands

You will need
- base ten blocks

GOAL

Relate thousands to hundreds and to tens.

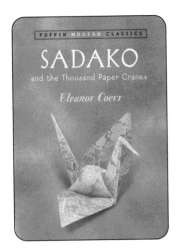

The students in Cory's school read *Sadako and the Thousand Paper Cranes* by Eleanor Coerr.
Each of the 350 students made 10 paper cranes.
They hung the cranes on strings in the gym.

? How many cranes did the 350 students make?

Cory's Model

I made 10 cranes.

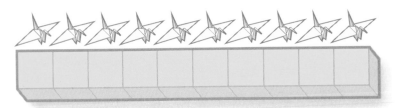

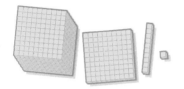

A. How can you model with one block the number of cranes made by 10 students?

B. How can you model with one block the number of cranes made by 100 students?

C. How many cranes did the 350 students make?

Reflecting

D. How does Part B show that 100 tens is 1000?

E. How does Part B show that 10 hundreds is 1000?

Checking

1. Elise's school has 420 students. Each student will make 10 cranes. How many hundreds of cranes will the students make?

Practising

2. In what other ways can you describe each number?
   a) 70 hundreds
   b) 20 hundreds

3. There are 100 cranes on each string. Represent the number of cranes as thousands and hundreds.
   a) 41 strings
   b) 17 strings

4. a) How many $100 bills would it take to pay for a piano that costs $9000? How do you know?
   b) How many $10 bills would it take to pay for the piano?

5. a) Explain how to describe 2000 as hundreds.
   b) When might you want to describe 2000 as hundreds instead of thousands?

Lesson 2

Place Value

You will need
- base ten blocks
- a place value chart

GOAL

Represent numbers to 10 000 using numerals, number words, and sketches.

Saddle Lake First Nation is in Alberta.
In 2006, the population of the community was 5742.
Another 2581 members of Saddle Lake First Nation lived outside the community.

Jade is sketching models of the population numbers.

 How can you represent the population numbers?

Jade's Model

For 5742, I say *five thousand seven hundred forty-two*.
I can write this on a place value chart.

Thousands	Hundreds	Tens	Ones
5	7	4	2

I can also **sketch** a model of this number.

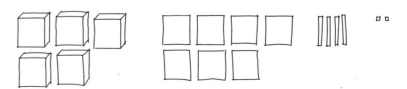

A. How do you know Jade's sketch represents 5742?

B. How many blocks did she use?

C. Sketch another model for 5742 using a different number of blocks.

D. Sketch a model for the 2581 members who live outside the community. Use the least number of blocks possible.

E. How many blocks did you sketch for Part D?

F. Sketch another model for 2581 using a different number of blocks.

Reflecting

G. How can you figure out the least number of blocks you need to model a 4-digit number with no zeros?

H. When you decide not to use the model with the least number of blocks, how many more blocks might you have to sketch? Explain.

Checking

1. There are 1943 members living in Big River First Nation in Saskatchewan.
   a) Sketch a model for 1943 using only hundreds, tens, and ones.
   b) Sketch a model for 1943 using the least number of blocks possible.

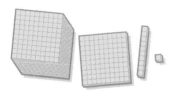

Practising

2. Sketch a model for each number using the least number of blocks possible.

   a) 1873   **b)** 6037   **c)** 4000   **d)** 6210

3. 3 spiders laid a total of 2065 eggs. Sketch 2 models for 2065.

4. Write the numeral for each number.

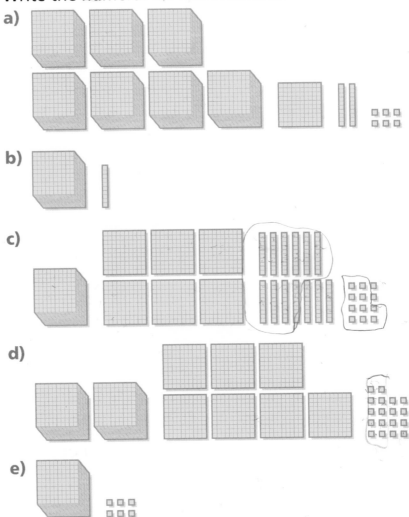

   a)

   b)

   c)

   d)

   e)

Reading Strategy

Compare your sketches of 2065 with a partner's. How are they the same?

5. Model each number with 12 base ten blocks. Sketch the model.

   a) 1803   **b)** 1002   **c)** 1020   **d)** 2001

6. A Chinese restaurant has sold 4562 spring rolls. It sells 100 more spring rolls each day.
   a) Sketch a model of 4562. Use the least number of blocks you can.
   b) For each of the next 5 days, add blocks to show the number of spring rolls sold. Make sure the model for each day uses the least number of blocks possible.
   c) What should the sign read at the end of 5 days?
   d) How did your sketch change for each day over the 5 days?

7. The number shown below is modelled with 5 thousands and some hundreds.

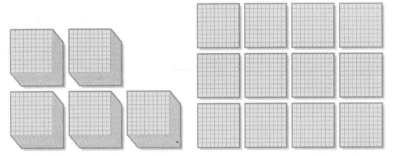

   a) What is the number?
   b) List 3 other numbers that can be modelled with 5 thousands and some hundreds.

8. Ian modelled a number with 10 hundreds blocks and some tens blocks. List 3 numbers that he might have modelled.

9. a) Sketch a model of 2222.
   b) What does each 2 represent?

10. a) Sketch 15 blocks to model a 4-digit number.
    b) List 4 more 4-digit numbers you can model with 15 blocks.

11. Why do you need at least 4 digits to represent a number greater than 999?

Lesson 3

Expanded Form

You will need
- place value charts
- counters

GOAL

Represent numbers to 10 000 using expanded form.

John Meisenheimer has a world-record collection of 4251 yo-yos.

How can you model 4251?

Emily's Models

I'll write 4251 on a place value chart.
Then I'll sketch a base ten block model.

Thousands	Hundreds	Tens	Ones
4	2	5	1

I can also use counters on the place value chart.

Thousands	Hundreds	Tens	Ones
4	2	5	1

Both of my models show 4 thousands, 2 hundreds, 5 tens, 1 one.

Luis's Representations

I can read 4251 as *four thousand two hundred fifty-one*.

When I write the number as 4251, it is in **standard form**.

I can also write 4251 in **expanded form** using words or numerals.

4 thousands + 2 hundreds + 5 tens + 1 one

4000 + 200 + 50 + 1

standard form

The usual way we write numbers

expanded form

A way to write numbers that shows the value of each digit

Reflecting

A. Why do you think Luis wrote 4251 as
4000 + 200 + 50 + 1
instead of as
200 + 4000 + 1 + 50?

B. Why does the expanded form for 4251 have more parts than the expanded form for 4002?

Communication Tip

There are 2 ways to write the standard form for 4-digit numbers. You can write 4567, or you can write 4 567, with a space between the thousands digit and the hundreds digit.

Checking

1. Daniel Evans collects computer mouse pads. He has 4567 mouse pads.
   a) How would you read 4567?
   b) Model 4567 on a place value chart.
   c) Represent 4567 as
      ▌ thousands + ▌ hundreds + ▌ tens + ▌ ones.
   d) Write 4567 in expanded form using numerals.

Practising

2. Rob Peterson holds a record with 8000 + 10 + 7 good tennis serves in a row.
   a) Write the number of serves in expanded form using words.
   b) Write the number of serves in standard form.

3. Write each number in standard form and in expanded form.

   a)
Thousands	Hundreds	Tens	Ones
● ●		● ● ● ● ●	

   b)
Thousands	Hundreds	Tens	Ones
● ● ●		●	● ● ● ●

4. a) Choose 2 four-digit numbers that can be modelled with 18 counters on a place value chart. Model each number.
   b) Write each number in expanded form.

5. Model each number on a place value chart. Write each in standard form.
   a) 6000 + 60 + 6
   b) 1 thousand + 2 hundreds + 1 ten

6. How many numbers have an expanded form that begins 2000 + 500 + ...? How do you know?

7. Explain how to write the expanded form of a 4-digit number in standard form. Use 3000 + 20 + 4 as an example.

Describing 10 000

GOAL

Explore and describe things involving 10 000.

Here are some ideas Ethan decided to use in his book about 10 000.

- A 10 000-word book is probably about 40 or 50 pages long.
- If you walk 10 000 steps, you can cross my bedroom 1000 times.
- 10 000 is the 10th number in the pattern 1000, 2000, 3000, ...

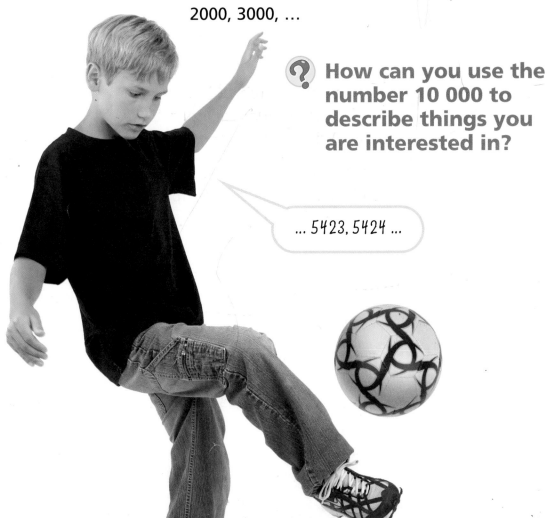

? **How can you use the number 10 000 to describe things you are interested in?**

... 5423, 5424 ...

Mid-Chapter Review

Frequently Asked Questions

Q: **How can you represent and describe a 4-digit number?**

A: You can use models, pictures, numbers, and words.

- standard form: 3205
- expanded form: 3 thousands + 2 hundeds + 5 ones or 3000 + 200 + 5
- words: three thousand two hundred five
- base ten blocks or pictures

Thousands	Hundreds	Tens	Ones

- counters or numbers

Thousands	Hundreds	Tens	Ones
3	2	0	5

Q: **What are some ways to represent 10 000?**

A: In words, it's ten thousand.
With blocks, it's 10 thousands blocks.
You can think of it as 1 more than 9999, 100 more than 9900, or 10 more than 9990.
It might be the number of people in a town.

Practice

Lesson 1

1. a) There are 43 bags with 100 candies in each. How many candies are there?
   b) If there were 10 candies in each bag, how many bags would be needed?

Lesson 2

2. A blue whale ate about 3750 kg of plankton. Sketch a model of 3750 using the least number of base ten blocks possible.

3. What number does each sketch represent?
   a)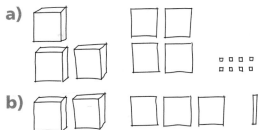

   b)

4. Show how to represent 1423 using each number of blocks.
   a) 10 blocks
   b) 19 blocks
   c) 28 blocks

Lesson 3

5. There are 4809 species of jumping spiders in the world. Write 4809 in expanded form.

6. Use standard form to write the number that is 3 thousands greater than one thousand twenty-nine.

7. Write each number in standard form.
   a) 2000 + 80 + 5
   b) 6 thousands + 2 hundreds + 5 tens + 6 ones

Writing Number Words

Writing Number Words

You will need
- Blank Cheques (blackline master)

GOAL

Write numbers to 10 000 using words.

The community centre bought a used car for $7795 to take seniors to appointments. They wrote a cheque to pay for the car.

 How can you write a cheque?

Cory's Cheque

I'll write the number in standard form on one line and in words on the other line.

The words have to tell the number of thousands, then the number of hundreds, and then the rest of the number.

```
                                                              656
XXX                                    Date ┌─┬─┬─┬─┬─┬─┬─┐
                                            └─┴─┴─┴─┴─┴─┴─┘
                                             D  D  M  M  Y  Y  Y

Pay to the order of _Smalltown Motors___    │ $ _7795_____ │

_Seven thousand seven hundred ninety-five_____  dollars

ꓚꓚꓚꓚ                              _____
```

Reflecting

A. How can you use expanded form to help you write the words for a 4-digit number?

Checking

1. Write a cheque for each amount.
   a) $9995
   b) $3950

2. Write the standard form for the amount of this cheque.

657

XXX   Date ☐ ☐ ☐ ☐ ☐ ☐ ☐ ☐
                                              D D M M Y Y Y Y

Pay to the order of _____ $ _____

Three thousand four hundred sixty
_____ dollars

0000   _____

Practising

3. Write a cheque for each amount.
   a) $1500
   b) $1005

4. Write each number in standard form.
   a) six thousand two hundred fourteen
   b) two thousand one hundred
   c) six thousand twenty-four

5. Rearrange these 6 words to describe 4 different numbers. Each time, write the number in standard form.

   six   hundred
   eight   two
   fifty thousand

6. Why do you think a bank needs both the standard form and the word form of the number on a cheque?

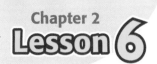
Locating Numbers on a Number Line

GOAL

Place 4-digit numbers on a number line.

Jade found some important dates in the history of British Columbia for her social studies project.

? **How can Jade place the dates on a timeline?**

1917
women won the
right to vote

2010
Olympic winter
games

1843
Victoria was
founded

1973
important
Aboriginal land
claim case

1778
Captain James
Cook landed on
Vancouver Island

Jade's Timeline

A timeline is just like a number line.
The numbers are the dates.
I'll start my line at 1700.
I'll count up by 50s.

1700 1750

A. Create a timeline that will allow you to include all 5 dates.

B. Place each date where it belongs on the line.

Reflecting

C. Why do you think Jade started her timeline at 1700?

D. Where did you end your timeline? Why did you end it there?

E. How did you know where to put each date?

Checking

1. Where would you put these 3 dates on your timeline? Explain your thinking.

1908
University of
British Columbia
founded

1872
British Columbia
joined Canada

1952
Voting age reduced
to 19

Practising

2. Estimate what number belongs at each letter.

a)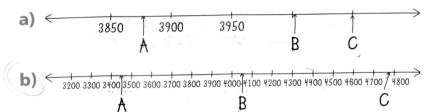

      3850   3900   3950   B   C
           A

b)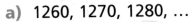

  3200 3300 3400 3500 3600 3700 3800 3900 4000 4100 4200 4300 4400 4500 4600 4700 4800
      A   B   C

3. Place the first 6 numbers of each pattern on a number line.

a) 1260, 1270, 1280, …

    1200 1210 1220

b) 2045, 3045, 4045, …

    2000 2500 3000

c) 2350, 2450, 2550, …

    2000 2100 2200

4. a) Create a number line to show counting by 100s from 1200 to 2300.

   b) Place each number on the line.

    2210   1823   1508   1298

5. What is wrong with each number line below? How do you know?

a)

    1400   1410   1420   1440   1450   1460

b)

    2900   3000   4000   5000   6000   7000

c)

    4000   4500   5000   6000   6500   7000

6. Why is skip counting important when you are creating a number line to model 4-digit numbers?

Movie Dates

Did you ever notice that at the very end of the credits for some movies there is a small note that contains letters like MMVI?

The letters are Roman numerals, and they tell the year the movie was made.

The chart shows some years using Roman numerals.

Roman Numerals

2004	2005	2006	2007	2008
MMIV	MMV	MMVI	MMVII	MMVIII

1. Where else have you seen Roman numerals?

2. What do you think the M in the Roman numerals stands for? Why do you think so?

3. What do you think the V stands for? Why?

4. The Romans used X for 10, L for 50, and C for 100. What do you think the Roman numerals for a movie made in each of these years would be?
   a) 2010
   b) 2011
   c) 2009

Comparing and Ordering Numbers

You will need
- place value charts
- counters

GOAL

Compare and order numbers to 10 000.

In 2005, *The Guinness Book of World Records* reported the greatest number of people who had ever got together at one time to pop balloons, throw snowballs, or skip rope.

 Were there more balloon poppers, snowball throwers, or rope skippers?

Ethan's Comparison

I'll show the numbers in a place value chart and on a number line.

Balloon Poppers

Thousands	Hundreds	Tens	Ones
1	6	0	3

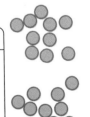

A. Finish modelling all 3 numbers on place value charts.

B. How can you use the place value charts to determine which number is the greatest?

C. Estimate to place 1356, 1603, and 2473 on a number line.

D. How can you use the number line to figure out which number is the least?

E. Compare the numbers using <, =, or >.
1356 ▨ 1603   2473 ▨ 1603   1356 ▨ 2473

Reflecting

F. Why was it easier to compare 2473 to 1603 than it was to compare 1356 to 1603?

G. Write a number between 1356 and 1603. What digit is in the thousands place? the hundreds place?

Checking

1. Look at the 2 world records.
   Which number is greater? Explain how you know.

2473 snowball throwers

5117 huggers

Practising

2. Compare each pair of numbers using <, =, or >.
   a) 1281 ▨ 654
   b) 6772 ▨ 7276
   c) 2395 ▨ 2942
   d) 1135 ▨ 1138

3. Replace ▨ with 3 different digits to make the sentence true. Explain how you know you're right.
   3▨25 > 3▨▨4

World Record Collections

Collection	Number of items
soft-drink cans	3284
car bumper stickers	3230
unused bandages	4500
yo-yos	4251
bottled water labels	5115
pencils	6885

4. Some world record collections are listed in the chart.
   a) Which collection has the fewest items? Explain how you know.
   b) Which collection has the most items? Explain how you know.

5. a) What is the least 4-digit number you can create using only the digits 7, 9, 7, and 0?
   b) What is the greatest 4-digit number?

6. What numbers are missing in each pattern?
   a) 4326, 4336, ▨, ▨, 4366
   b) 7216, ▨, ▨, ▨, 7616, 7716

7. Explain how you can use the digits of 2 four-digit numbers to decide which number is greater.

8. a) Use the digits 0 to 9 to create 3 different 4-digit numbers. Don't use any digit more than twice.
   b) Order the numbers from greatest to least.

9. How is the process for comparing 4-digit numbers the same as the process for comparing 3-digit numbers?

Target 3000

Number of players: whole class, small
group, or pairs

How to play: Make 4-digit numbers from
rolls of a die. Make numbers as
close to 3000 as possible.

- **Step 1** Each player rolls a die 4 times.

- **Step 2** The player records the digits in the form .
  The player can record the digits in any order.

- **Step 3** The player closest to 3000 wins a point.

The first player with 10 points wins the game.

Luis's Turn

Cory rolled 5, 1, 2, 4.
His best number is 2541.

I rolled 3, 3, 2, 1.
My best number is 3123.

My number is closer to 3000.
I win a point.

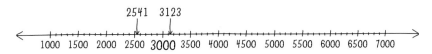

Communicating about Ordering Numbers

GOAL

Explain how to order a set of numbers in a complete, clear, and organized way.

Top Scores
909
9009
999
9909

Emily organized a computer-game contest for her sister's friends.

When she awarded 1st, 2nd, 3rd, and 4th prizes, the players asked how she ordered the scores.

 How can Emily explain the steps she followed to order the numbers?

Emily's Explanation

Good Copy

The winning scores were 909, 9009, 999, and 9909.

1. I wrote the numbers with 4 digits first. They are all greater than 3-digit numbers since they are all more than 1000.
2. 9909 is greater since it has 9 hundreds and 9009 has no hundreds.
3. Then I wrote the numbers with 3 digits.
4. I compared the tens digits to see which number was greater.
   999 > 909 since it has 9 tens and 909 has no tens.
   The order is 9909 9009 999 909.

Rough Copy

I wrote the numbers with 4 digits first.
I wrote the numbers with 3 digits last.
I compared the digits to put them in order.

58

Communication Checklist

✔ Did you show all the steps?
✔ Did you explain your thinking?

Reflecting

A. Describe some differences between Emily's rough copy and her good copy. Use the Communication Checklist.

Checking

1. Cory put these game scores in order from greatest to least.

865　1876　1540　86　1000

Here is the rough copy explaining his steps. Use the Communication Checklist to write a good copy.

> *Rough Copy*
>
> I wrote 1000 in the middle.
> I wrote the 4-digit numbers first.
> I wrote the numbers less than 1000 last.
> 1876　1540　1000　86　865
> They were all in order except 86 and 865, so I switched them.
> 1876　1540　1000　865　86

Practising

2. a) Order these numbers from greatest to least.
   3867　3869　392　473　450
   b) Write an explanation of how you did the ordering. This is your rough copy.
   c) Use the Communication Checklist to find ways to improve your rough copy. Then write a good copy.

Chapter Review

Frequently Asked Questions

Q: **How can you order 3 four-digit numbers?**

A1: You can compare them 2 at a time using a place value chart. Compare thousands first, and, if necessary, hundreds, then tens, then ones. For example, order 4123, 4089, and 4312 from least to greatest.

Thousands	Hundreds	Tens	Ones	
●●●●	●	●●	●●●	4123
●●●●	●●●	●	●●	4312

4312 and 4123 both have 4 thousands, but 4312 has more hundreds.

Thousands	Hundreds	Tens	Ones	
●●●●	●	●●	●●●	4123
●●●●		●●●●●●●●	●●●●●●●●●	4089

4123 has more hundreds than 4089.
4089 < 4123 < 4312

A2: You can compare all 3 numbers using a number line.

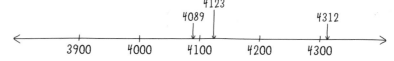

4312 is the greatest because it is farthest to the right. 4089 is the least because it is farthest to the left. 4123 is between the other two.
4089 < 4123 < 4312

Practice

Lesson 1

1. How many $100 bills would you need to pay for each?

a)  $3000   b) $2500   c) $3300

2. How many $10 bills would you need to pay for each?

a)  $2000   b) $330   c) $3300

Lesson 2

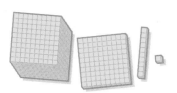

3. Sketch base ten blocks to model the number two thousand nine.

4. Write the numeral for this number.

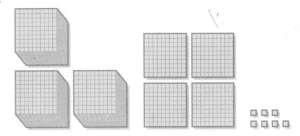

Lesson 3

5. There are 2230 organic farms in Canada.
   a) Model 2230 with the least number of blocks possible. Sketch the blocks.
   b) Write 2230 in expanded form using numbers and using words.

6. Write each number in standard form.
   a) 1000 + 90 + 6
   b) six thousand one hundred twenty-nine

Lesson 4

7. Describe the number 10 000 in 2 ways.

8. Write each number in words.
   a) 3105   b) 8002

9. Draw a number line and estimate the location of 2004, 3520, and 2879.

10. Which numbers are missing in each pattern?
    a) 2340, 2350, 2360, _____, _____, 2390, _____
    b) 4235, 4245, 4255, _____, _____, 4285

11. Which number on each number line is misplaced?

    a)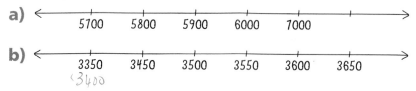

    b)
    3350   3450   3500   3550   3600   3650
    ‹3400›

12. Copy the number sentence below 3 times. Place different digits in the boxes each time to make it true. Explain your thinking.
    ▧ 295 > 15 ▧ 4

13. Maddy uses a pedometer to measure the number of steps she walks each day. Order the days from the least number of steps taken to the most.

Steps Taken

Wed.	Thurs.	Fri.	Sat.	Sun.
6214	7158	6043	8124	7053

What Do You Think Now?

Look back at **What Do You Think?** on page 35. How have your answers and explanations changed?

Chapter Task

Creating a Puzzle

Hugo
The population is the greatest number you can make that is less than 10 000.

Centuria
The population can be described as a number of hundreds.

Octavia
The population is a bit more than 8000 and can be modelled using only thousands and tens.

Middletown
The population is between 1200 and 1300 and can be modelled with only hundreds and tens.

Emily created a population puzzle with 4 make-believe towns.

- All of the towns have 4-digit populations.
- Each population can be modelled using 16 base ten blocks.

? **What is the population of each town?**

A. Use Emily's clues to figure out the population of each town. Explain your thinking for each answer.

B. List the populations of the towns in order. Explain how you ordered the populations.

C. Name a town with a population between 1000 and 10 000 in your province or territory. Make up a clue so that someone can figure out the population.

Chapter 3

Addition and Subtraction

GOALS

You will be able to

- estimate sums and differences
- add and subtract with 3-digit numbers and 4-digit numbers in different ways
- create and solve addition and subtraction problems
- communicate about estimation strategies

Create one addition problem and one subtraction problem using the picture on these pages. Solve your problems.

Getting Started

Counting Students

The bar graph shows the numbers of students who go to 3 different schools.

North School has only 25 students.

? **About how many students go to the 3 schools?**

Elementary School Students

North School

Lakeview School

South School

Number of students

A. Estimate the number of students at South School. Explain how you estimated.

B. Estimate the number of students at Lakeview School. Explain how you estimated.

C. Estimate the total number of students at the 3 schools.

D. Suppose North School had 50 students. How would your answers to Parts A to C change?

What Do You Think?

Do you *agree* or *disagree* with each statement?
Explain your thinking.

1. You can use 20 + 20 = 40 to calculate 18 + 19 and 40 − 19.

2. If you cut 41 cm from a 100 cm roll of ribbon, you will have more than 60 cm left.

3. If you add a number in the 300s to a number in the 500s, the answer is always in the 800s.

4. There is more than one way to subtract 99 from 200.

Solving Problems by Estimating

You will need
- base ten blocks
- a place value chart

GOAL

Estimate sums of 3-digit numbers to solve problems.

Each student in Lang's class is writing a story that has more than 500 words. Lang wrote a 3-page story.

My Riding Adventure

Page 1
245 words

Page 2
235 words

Page 3
135 words

 Did Lang write more than 500 words?

Lang's Solution

I'll use base ten blocks to show the number of words on each page.

Hundreds	Tens	Ones	
			245
			235
			135

I don't need to calculate an exact answer. I can estimate. I'll think about the hundreds first.

A. How can you use mental math to add the hundreds?

B. How can adding the hundreds help you to estimate the sum?

C. Did Lang write more than 500 words? Explain how you know.

Reflecting

D. Why didn't Lang need to calculate an exact answer?

E. Suppose his last page had 35 words instead of 135 words. Did Lang write more than 500 words? Explain how you know.

Checking

1. Maya wrote 275 words on one page of her story. She wrote 250 on the other. Does her story have more than 500 words? Would you estimate or would you calculate an exact answer? Explain.

Practising

2. Estimate. Explain how you estimated one sum.
   a) $567 + 813$ b) $611 + 149$

3. Use mental math to calculate the top sum. Use that sum to estimate the sum below.
   a) $200 + 150 = $ ▢   b) $600 + 200 + 200 = $ ▢
      $198 + 152 = $ ▢ $578 + 207 + 195 = $ ▢

4. A school raised $800 to buy a computer and printer like the ones shown. Does the school have enough money to buy both? Explain how you know.

5. Why is it important to know how to estimate the sum of 2 or more 3-digit numbers?

Estimating Sums

You will need
- base ten blocks
- a place value chart
- number lines

GOAL

Estimate sums in different ways.

Aneela plans to run this route along the Meewasin Trail in Saskatoon.

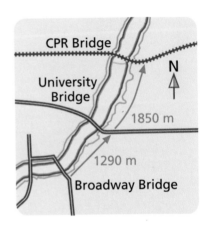

CPR Bridge
University Bridge
N
1850 m
1290 m
Broadway Bridge

 About how far will Aneela run?

Kate's Estimate

I'll estimate 1290 m + 1850 m using base ten blocks.

Thousands	Hundreds	Tens	Ones

I'll just add the thousands.
1 thousand + 1 thousand = 2 thousand
1000 + 1000 = 2000
Aneela will run more than 2000 m.

Cole's Estimate

I'll estimate 1290 + 1850 by adding the closer thousands.

1290 is closer to 1000 than 2000.
1850 is closer to 2000 than 1000.
1 thousand + 2 thousand = 3 thousand

Aneela will run about 3000 m.

Joshua's Estimate

I'll use a number line to estimate 1290 + 1850.
I'll start at 1300 and add 1000 and then 800.

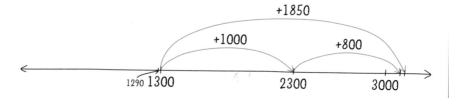

1300 + 1000 = 2300
2300 + 800 is greater than 3000.

Aneela will run more than 3000 m.

Reflecting

A. How can you improve Kate's estimate?

B. How do you think Joshua knows that 2300 + 800 is greater than 3000?

Checking

1. Kate plans to run this route over the 2 bridges. Use 2 ways to estimate how far she will run. Explain each estimate.

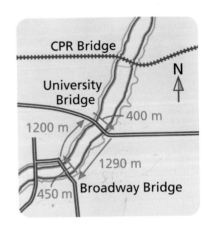

Practising

2. Estimate. Explain 2 of your estimates.
   a) 567 + 513
   b) 4235 + 915
   c) 4611 + 3449
   d) 1436 + 4602 + 997

3. How do you know each statement is true?
   a) The answer to 2867 + 4189 is between 6000 and 8000.
   b) The sum of 2867 and 4189 is close to 7000.

4. The chart shows the number of people who went to a 3-day folk festival. Estimate the total attendance.

Folk Festival Attendance

Day	Attendance
Thursday	899
Friday	1799
Saturday	2375

5. There is only one correct answer when you calculate a sum. Use 2 four-digit numbers to show that there can be more than one good estimate of a sum.

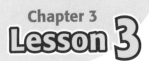

Exploring Addition and Subtraction

> **GOAL**
>
> Use your own strategies to add and subtract numbers to solve a problem.

Jade made jingle dresses for a powwow with her mother and younger sister. They folded 1000 pieces of metal into the shape of cones. They sewed the cones onto 3 dresses.

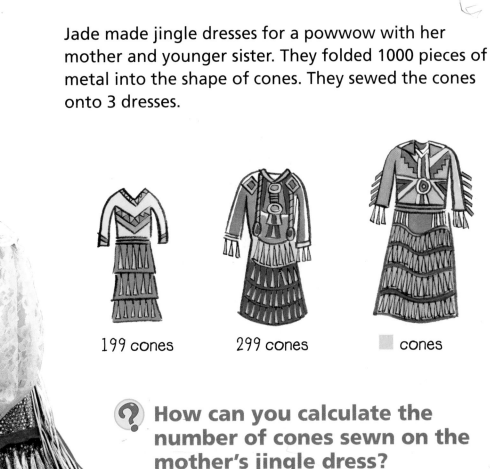

199 cones   299 cones   ▨ cones

? How can you calculate the number of cones sewn on the mother's jingle dress?

Adding from Left to Right

You will need
• grid paper

> **GOAL**
>
> Solve addition problems by adding from left to right.

A forklift operator wants to lift 3 containers together. The forklift can safely lift up to 4000 kg.

1589 kg Bottled Water

719 kg Soup Cans

1632 kg Juice Boxes

? **Can the forklift safely lift all 3 containers together?**

Tien's Addition

I predict that the total mass is close to 4000 kg.
I'll calculate an exact answer.
I'll add the numbers in the place value columns from left to right.

- **Step 1** First, I'll add the thousands.

```
    1 5 8 9
      7 1 9
  + 1 6 3 2
    2 0 0 0
```

- **Step 2** Next, I'll add the hundreds.

```
    1 5 8 9
      7 1 9
  + 1 6 3 2
    2 0 0 0
    1 8 0 0
```

A. Explain why Tien wrote 1800 in Step 2.

B. What sum do you think she will record next?

C. Complete Tien's addition. Show your work.

D. Can the forklift safely lift the 3 containers together? How do you know?

Reflecting

E. How do you think Tien predicted that the total mass of the 3 containers was close to 4000 kg?

F. Why does adding from the left give you a good estimate for your answer?

2455 kg Books
849 kg DVDs
4567 kg Batteries

Checking

1. Another forklift can lift 8000 kg safely.
   a) Can it lift these 3 containers together?
   b) Did you estimate to solve the problem or did you calculate an exact answer? Explain.

Practising

2. Estimate each sum. If your estimate is less than 9000, calculate an exact answer.
   a) 4273 + 3935
   b) 5657 + 3456
   c) 2456 + 6514
   d) 4689 + 4291 + 867
   e) 2599 + 2466 + 1985
   f) 1578 + 3355 + 4012

3. Three schools recycled telephone books to raise money.
   a) How many telephone books did they recycle altogether?
   b) Is your answer reasonable? How do you know?

4. Cole and Julia each calculated 1777 + 789.

 Cole's Calculation

```
  1777
+  789
  1000
  1400
   150
 +  16
  2566
```

Julia's Calculation

```
  1777
+  789
  2400
   150
 +  16
  2566
```

   a) Why do you think Cole added 4 numbers and Julia added only 3 numbers?
   b) How might Cole and Julia add 3882 + 938?

5. When you add a 4-digit number to a 3-digit number, how do you know which digits to add? Use 2568 + 987 to help you explain.

6. Calculate. Show your work.
   a) 1259 + 618
   b) 6963 + 2364
   c) 4211 + 345 + 967
   d) 1567 + 1578 + 2567

7. How is adding 2 four-digit numbers similar to adding 2 three-digit numbers? How is it different?

76

Race to 1500

Number of players: 2
How to play: Cross out numbers and calculate the total.

You will need
- grid paper
- pencil crayons

- **Step 1** Copy this game card.

- **Step 2** Player 1 crosses out a number.

- **Step 3** Player 2 crosses out another number using a different colour. Player 2 calculates the sum of the crossed-out numbers.

50	100	150	200	250	300
50	100	150	200	250	300
50	100	150	200	250	300
50	100	150	200	250	300

- **Step 4** Players take turns crossing out and adding all the numbers.

The player who reaches a total of exactly 1500 wins.

Lang's Turn

Kate and I have crossed out a total of 1200 so far. I can win if I cross out a 300!

50	~~100~~	150	200	250	~~300~~
50	~~100~~	150	200	~~250~~	~~300~~
50	100	~~150~~	200	250	300
50	100	150	200	250	300

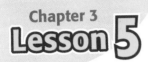
Adding from Right to Left

You will need
- base ten blocks
- a place value chart
- grid paper

Solve addition problems by adding from right to left.

Aneela's school collects food labels to get points. They can trade the points for school equipment.

(?) How many points does Aneela's school need for the bag and glove?

Aneela's Addition

I need to calculate 3355 + 2745.

I predict that we'll need more than 6000 points.

- **Step 1** I'll calculate the actual number of points by adding base ten blocks on a place value chart. I'll record on grid paper.

Thousands	Hundreds	Tens	Ones

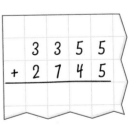

$$\begin{array}{r} 3\ 3\ 5\ 5 \\ +\ 2\ 7\ 4\ 5 \\ \hline \end{array}$$

regroup

Trade 10 smaller units for 1 larger unit, or 1 larger unit for 10 smaller units

- **Step 2** I'll add the ones. Then I'll **regroup**.

  5 ones + 5 ones = 10 ones

  10 ones = 1 ten + 0 ones

Thousands	Hundreds	Tens	Ones

		1		
	3	3	5	5
+	2	7	4	5
				0

- **Step 3** I'll add the tens.

  1 ten + 5 tens + 4 tens = 10 tens

  10 tens = 1 hundred + 0 tens

Thousands	Hundreds	Tens	Ones

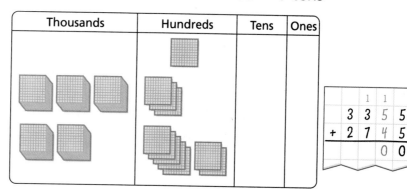

		1	1	
	3	3	5	5
+	2	7	4	5
			0	0

A. Complete Aneela's addition. Show your work.

B. How many points does the school need for the bag and glove? How do you know your answer is reasonable?

Reflecting

C. How can you tell when you need to regroup when adding from right to left? Use an example to help you explain.

Checking

1. Calculate the number of points the school needs to get all 3 books. Is your answer reasonable? Explain.

Points Needed

Book	Number of points needed
Sports book	1825
Astronomy book	1175
Dinosaur book	825

Practising

2. An online discussion group has a goal of 7500 postings. In January, it had 1535 postings. In February, it had 2865 postings. In March, it had 3145 postings. Did the group reach its goal? Explain.

3. Estimate each sum. If your estimate is between 4000 and 6000, calculate the exact answer.
   a) 2987 + 145
   b) 3254 + 2162
   c) 2311 + 2499
   d) 2300 + 2253 + 1701

4. You can arrange these 7 cards to form a 3-digit number and a 4-digit number.

   a) What is the greatest sum you can create?
   b) What is the least sum you can create?

5. Why is it useful to know more than one way to add 2 four-digit numbers?

Mid-Chapter Review

Frequently Asked Questions

Q: **How can you decide whether to estimate or calculate to solve a problem?**

A: You can estimate when you don't need an exact answer. For example, suppose you want to walk at least 8000 steps on the weekend.
You walk 3487 steps on Saturday and 4675 steps on Sunday.
You can estimate 3487 + 4675 by adding thousands and hundreds.

$$3000 + 4000 = 7000$$
$$400 + \ 600 = 1000$$
$$7000 + 1000 = 8000$$

You're already at 8000 and still have more to add. You don't need to calculate any more.

Q: **How can you add 3-digit and 4-digit numbers?**

A: You can use number lines, mental math, base ten blocks, or methods like these:

You can add from left to right.

```
  1689
+  356
  1900
   130
 +  15
  2045
```

You can add from right to left and regroup when needed.

```
    1 1 1
    1 6 8 9
+     3 5 6
    2 0 4 5
```

Practice

Lesson 1

1. A food bank wants to raise $8000. It has $5473 and receives a donation of $2750.
   a) Has the food bank reached its goal?
   b) Did you estimate or did you calculate an exact answer to solve the problem? Explain.

Lesson 2

2. One answer for each calculation is correct. Estimate to identify the correct answer.
   a) $2367 + 2710 = $ ▮   5077 or 6077
   b) $2986 + 145 + 3978 = $ ▮   6109 or 7109

Lesson 4

3. Calculate. Show your work.
   a) $4567 + 97$ b) $238 + 438 + 498$

Baseball Attendance

	Day 1	Day 2
2006	1345	2267
2007	1456	2039

4. A 2-day baseball tournament is held annually. The chart shows the attendance in 2006 and 2007.
   a) Which year had the greater attendance?
   b) Did you estimate or did you calculate an exact answer? Explain.

Lesson 5

5. Would you calculate each sum using mental math or using pencil and paper? Give a reason for each choice. Then write the sum.
   a) $26 + 24$ c) $7764 + 1495$
   b) $3200 + 1800$ d) $3855 + 3755$

Movie Attendance

Day	Attendance
1	1676
2	2659
3	3189

6. The chart shows the number of people who saw a movie in its first 3 days in a theatre.
   a) How many people saw the movie?
   b) Explain how you know your answer is reasonable.

7. Calculate. Show your work.
   a) $5679 + 345$ b) $5432 + 986 + 238$

Lesson 6 Estimating Differences

GOAL

Use your own strategies to estimate differences.

Tien made a chart that shows the highest peaks in Western and Northern Canada. She wants to know about how much taller Mount Everest is than the highest peaks in her chart.

Highest Peaks in Western and Northern Canada

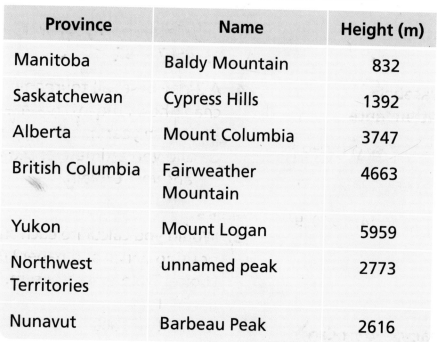

Province	Name	Height (m)
Manitoba	Baldy Mountain	832
Saskatchewan	Cypress Hills	1392
Alberta	Mount Columbia	3747
British Columbia	Fairweather Mountain	4663
Yukon	Mount Logan	5959
Northwest Territories	unnamed peak	2773
Nunavut	Barbeau Peak	2616

? How can you estimate the difference between the height of Mount Everest and the height of each peak in Tien's chart?

Subtracting Numbers Close to Hundreds or Thousands

You will need
- number lines

GOAL

Use mental math to subtract.

In 1899, Treaty No. 8 was signed by the Cree, Beaver, and Chipewyan First Nations and the Government of Canada. The treaty covered parts of what are now British Columbia, Alberta, Saskatchewan, and the Northwest Territories.

How many years have passed since the treaty was signed?

Joshua's Method

I'll use the year 2007. I can show 1899 and 2007 on a number line.

1899 + ▧ = 2007
2007 − 1899 = ▧

I can add on to 1899 to get to 2007.

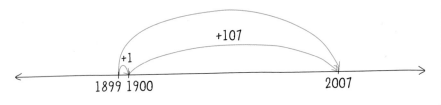

Kate's Method

I'll show the 2 dates on a number line.

1899   2007

If I add 1 year to each date, the difference between the 2 dates will be the same, but the new numbers will be easier to subtract.

+1   +1

1900   2008

Then 2007 − 1899 becomes 2008 − 1900.

A. How do you know that 2007 − 1899 is the same as 2008 − 1900?

B. Complete the subtraction using Joshua's method and then Kate's method. Explain what you did.

C. How many years have passed between 1899 and 2007?

Reflecting

D. Why do you think Joshua decided to add the 1 first?

E. How did you use mental math to subtract 1900 from 2008 when you completed Kate's method?

Checking

1. In 1792, Captain George Vancouver was the first European to explore Burrard Inlet, where the city of Vancouver is now. In 2010, the Winter Olympics will be held in Vancouver. How many years are between the 2 dates? Explain how you found out.

Practising

2. Calculate the difference between your age and the age of an adult you know. How do you know that the difference between your age and the adult's age will always be the same, even as you both get older?

3. Use mental math to calculate. Explain what you did for 2 answers.
   a) $200 - 98$
   b) $1000 - 298$
   c) $5000 - 1998$
   d) $2007 - 999$

4. The mass of an empty helicopter is 2998 kg. When it is loaded, the helicopter can have a maximum mass of 4536 kg. What is the difference between its empty mass and its maximum mass?

5. Show how to change some digits in each number so that the difference will still be 1671.

	3	3	6	0
−	1	6	8	9
	1	6	7	1

6. Calculate.
   a) $317 + \blacksquare = 2010$
   b) $2998 + \blacksquare = 4000$

7. Vera thinks it's easy to subtract a number like 999 from a 4-digit number. Use an example to help you explain why she might think so.

Subtracting Another Way

Before he came to Canada, Khaled learned this method for subtracting.

- **Step 1**   To subtract 567 from 1234, first I add 10 ones (10) to the top number and 1 ten (10) to the bottom number. Then I can subtract 7 ones from 14 ones.

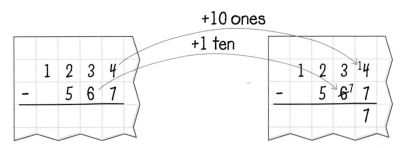

- **Step 2**   Next I add 10 tens (100) to the top number and 1 hundred (100) to the bottom number. Then I can subtract 7 tens from 13 tens.

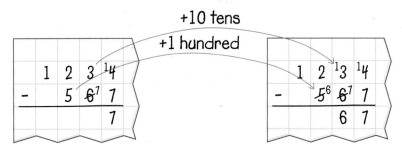

1. How do you know that adding 10 tens to the top number and 1 hundred to the bottom number will not change the difference?

2. Complete Khaled's method.

3. Use Khaled's method to calculate each difference.
   a) 3000 − 356   b) 5600 − 1453

Chapter 3
Lesson 8

Regrouping before Subtracting

You will need
- a place value chart
- base ten blocks
- grid paper

 GOAL

Solve subtraction problems by regrouping first.

A video store has 1257 DVDs.
848 DVDs have already been rented.

? **How many DVDs are left at the video store?**

Julia's Subtraction

I think 1257 − 848 must be about 400 because
12 hundreds − 8 hundreds = 4 hundreds.
I can subtract to calculate the actual difference.

- **Step 1** I'll compare the numbers column by column.
  I need more than 2 hundreds to take away 8 hundreds.

Thousands	Hundreds	Tens	Ones

$$\begin{array}{r} 1\ 2\ 5\ 7 \\ -\quad 8\ 4\ 8 \\ \hline \end{array}$$

- **Step 2** I can regroup 1 thousand 2 hundreds as 12 hundreds.

Thousands	Hundreds	Tens	Ones

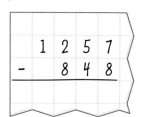

- **Step 3** I still need more ones before I can take away 8 ones.
  I'll regroup 5 tens 7 ones as 4 tens 17 ones.

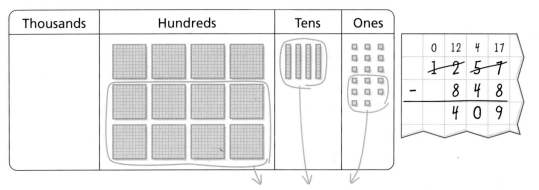

Thousands	Hundreds	Tens	Ones

```
  0  12  4  17
  1  2   5  7
-     8  4  8
```

- **Step 4** Now I have enough in each place value column to take
  away 848.

Thousands	Hundreds	Tens	Ones

```
  0  12  4  17
  1  2   5  7
-     8  4  8
      4  0  9
```

There are 409 DVDs left at the video store. This answer seems
reasonable because it's pretty close to my estimate of 400.
I can also add to check: 409 + 848 = 1257.

Reflecting

A. In Step 2, why do you think Julia regrouped
1 thousand 2 hundreds as 12 hundreds?

B. How does Step 4 show that 1257 was
regrouped as 1200 + 40 + 17?

Checking

1. 788 of the 1257 DVDs had been rented.
   a) How many DVDs were left at the video store?
   b) Explain how you know your answer is reasonable.

Practising

2. Calculate. Check by adding.
   a) 3469 − 278   b) 1723 − 1576

3. The video store was giving away 4000 discount coupons. By Wednesday, 1571 coupons had been given away. How many coupons were left?

4. Estimate each difference. If your estimate is greater than 2000, calculate an exact answer.
   a) 2348 − 999   b) 2749 − 682

5. Would you calculate each difference using mental math or using pencil and paper? Give a reason for each choice. Then calculate.
   a) 5324 − 324   c) 1542 − 500
   b) 6905 − 2876   d) 9000 − 6999

6. What is the greatest difference you can get when you subtract a 3-digit number from a 4-digit number? Explain how you know.

7. Jon said you can subtract from left to right.

   $$\begin{array}{r} 2000 \\ -\ 575 \\ \hline \cancel{1500} \\ 1425 \end{array}$$
   2000 − 500 = 1500
   1500 − 75 = 1425

   a) Explain how his method works.
   b) Use his method to calculate 5000 − 495.

8. When you subtract a 3-digit number from a 4-digit number, how do you know which digits to regroup? Use 2568 − 917 to help you explain.

Target 3500

You will need
- a die

Number of players: 2 to 4
How to play: Players estimate and calculate differences.

- **Step 1** A player rolls a die 4 times and uses the numbers to write a 4-digit number.

- **Step 2** Each player estimates the difference between the number and the target, 3500.
  Players record their estimates.

- **Step 3** All players calculate the difference.

- **Step 4** Each player scores points for his or her estimate:
  Estimate within 500: 1 point
  Estimate within 100: 2 points
  Estimate within 50: 3 points

- **Step 5** Continue for 5 turns.

The player with the greatest number of points wins.

Aneela's Turn

I rolled 1262. I estimate that 3500 − 1262 is close to 2300. My estimate of 2300 is within 100 of the answer of 2238. I score 2 points.

$$\begin{array}{r} \overset{9}{}\overset{10}{} \\ \overset{4}{3}\ \overset{10}{\cancel{5}}\ \cancel{0}\ \cancel{0} \\ -\ 1\ 2\ 6\ 2 \\ \hline 2\ 2\ 3\ 8 \end{array}$$

Subtracting by Renaming

GOAL

Use renaming to make subtraction easier.

Vera calculated that she had lived 3286 days. She wanted to know the number of days until her 5000th day birthday.

? **How many days are there until Vera's 5000th day birthday?**

Cole's Subtraction

I estimate that 5000 − 3286 is close to 5000 − 3300. I think it's about 1700 days.
To calculate 5000 − 3286, I'll rename 5000 so it's easier to subtract.

- **Step 1** I can rename 5000 as 4999 + 1.

  $$
  \begin{array}{r}
  4999 + 1 \\
  \cancel{5000} \\
  - 3286 \\
  \end{array}
  $$

- **Step 2** Now I can subtract 3286.

  $$
  \begin{array}{r}
  4999 + 1 \\
  \cancel{5000} \\
  - 3286 \\
  \hline
  1713 + 1 = 1714 \\
  \end{array}
  $$

Vera's 5000th day birthday is in 1714 days.

Reflecting

A. Why do you think Cole added 1 to 1713 in Step 2?

B. How is Cole's method like other subtraction methods? How is it different?

Checking

1. Vera's brother is 1083 days old. How many days are there until his 5000th day birthday?

Practising

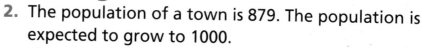

2. The population of a town is 879. The population is expected to grow to 1000.
  a) By how much will the population grow?
  b) How do you know your answer is reasonable?

3. Kyle has 3456 points in a game. To win, he must score 6000 points. How many more points does he need to win?

4. Estimate. Then calculate.
  a) 1000 − 435
  b) 2000 − 435
  c) 3000 − 278
  d) 6000 − 332

5. In a town of 7000 people, 914 of the people are 6 years old or younger. How many people are older than 6 years?

6. Show 2 ways to calculate 1000 − 250. Which way was easier for you? Explain.

7. Jack subtracted by renaming 3000 as 2999 + 1.
  a) What step is missing?
  b) What is another way Jack could rename 3000?

$$\begin{array}{r} 2999 + 1 \\ \cancel{3000} \\ -\ 437 \\ \hline 2562 \end{array}$$

Communicating about Number Concepts and Procedures

> **GOAL**
>
> Explain your thinking when estimating a sum or difference.

There were 9000 tickets for sale for a hockey game at Memorial Arena. 2488 tickets for blue seats were sold. 4934 tickets for green seats were sold. Joshua calculated that 1578 tickets were unsold.

 How can you explain whether Joshua's calculation is reasonable?

Tien's Explanation

I'll estimate to check Joshua's calculation, but I have to explain my steps.

2488 is about 2500.

4934 is about 5000.

$2500 + 5000 = 7500$

About 7500 tickets were sold.

$9000 - 7500 = 1500$

About 1500 tickets were unsold.
So 1578 is a reasonable answer.

Lang's Explanation

I'll subtract 2488 and 4934 from 9000 in 2 steps to check Joshua's answer of 1578.

9000 − 2488 is about 9000 − 2500 = 6500.

6500 − 4934 is about 6500 − 5000 = 1500.

So an answer of 1578 is reasonable.

Communication Checklist

✔ Did you show the right amount of detail?

✔ Did you explain your thinking?

Reflecting

A. How do you think Tien and Lang would answer each question in this Communication Checklist? Why?

Checking

1. The arena has 6000 tickets for the circus. 1631 adult tickets and 3712 children's tickets were sold. How many tickets were unsold? Explain how you know your answer is reasonable.

Practising

2. Bryan scored 2815 points and 3947 points in levels 1 and 2 of a video game. How many points does he need to score in level 3 to reach 7500 points? Explain how you know your answer is reasonable.

3. A biologist recorded a total of 3226 salmon spawning in 3 creeks. How many salmon were spawning in Bent Creek? Explain how you know your answer is reasonable.

4. Why is it important to explain your thinking when estimating a sum or difference?

Salmon Spawning

Creek	Number
Stoney Creek	1678
Swift Creek	749
Bent Creek	

Chapter Review

Frequently Asked Questions

Q: How can you estimate a difference?

A: You can use the closest thousands or hundreds and use mental math to subtract.
For example, subtract 1876 from 8000. You can estimate in steps.
$8000 - 1000 = 7000$
$7000 - 876$ is about the same as
$7000 - 900 = 6100$.
So you can estimate that $8000 - 1876$ is about 6100.

Q: How can you subtract 3-digit and 4-digit numbers?

A1: You can add on from the lesser number to the greater number.
For example, calculate $2000 - 975$.

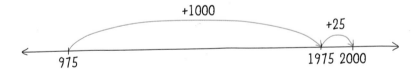

$1000 + 25 = 1025$
So $2000 - 975 = 1025$.

A2: You can use regrouping.
For example, calculate
$4793 - 889$.

	3	17	8	13
	4	7	9	3
−		8	8	9
	3	9	0	4

Practice

Reading Strategy

Choose one new idea you learned in this chapter. Show what you know about this idea.

Lesson 1

1. One answer for each calculation is correct. Estimate to identify the correct answer.

a) $159 + 298 = \square$ 457 or 557

b) $318 + 578 = \square$ 796 or 896

c) $198 + 289 + 358 = \square$ 745 or 845

2. a) Will the total of these 3 lengths be greater than 3000 cm?

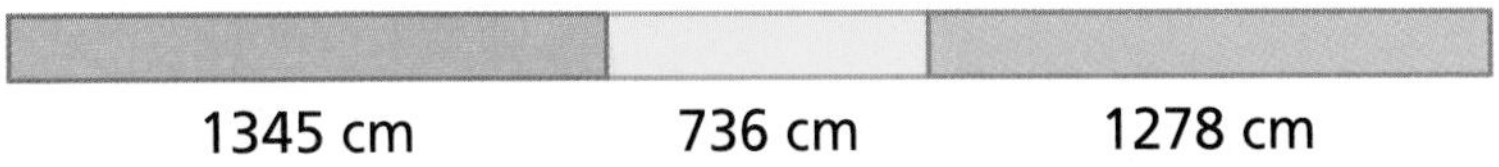
1345 cm
736 cm
1278 cm

b) Did you estimate or did you calculate an exact answer? Explain.

Lesson 2

3. Annik wants to walk around one of these parks. Which walk would be longer? Explain how you know.

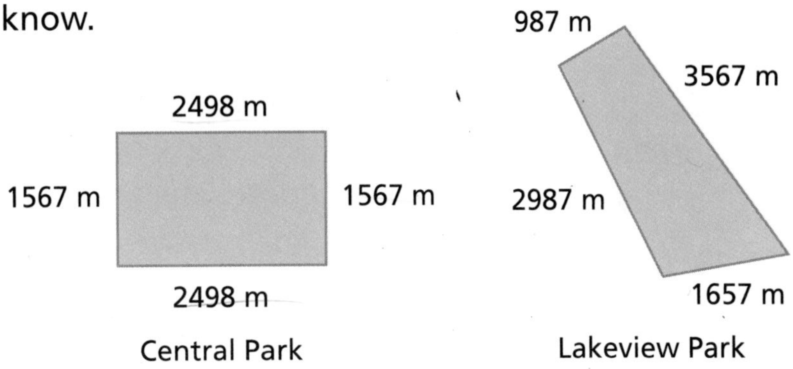
2498 m
1567 m
1567 m
2498 m
Central Park
987 m
3567 m
2987 m
1657 m
Lakeview Park

Lesson 4

4. An online store sold 3435 books on Saturday and 3675 books on Sunday. How many books were sold in total?

Lesson 5

5. A plane is travelling 2567 m higher than a mountain that is 1868 m high.

a) Estimate how high the plane is flying.

b) Calculate how high the plane is flying.

6. Calculate. Show your work.
   a) 1249 + 458
   b) 236 + 7638
   c) 1387 + 6729
   d) 6465 + 1878

Lesson 7

7. Use mental math to subtract. Explain what you did.
   a) 230 − 99
   b) 600 − 375
   c) 1000 − 299
   d) 6000 − 1998

Lesson 8

8. A truck carrying 2 crates has a total mass of 8341 kg. One crate has a mass of 1499 kg. The truck by itself has a mass of 5443 kg. What is the mass of the other crate?

9. Estimate. Then calculate.
   a) 4376 − 257 = ▩
   b) 3695 − 776 = ▩
   c) 2371 − 640 = ▩
   d) 9018 − 1766 = ▩

Lesson 9

10. Pedro plans to walk 5000 steps in one day. His pedometer shows he has walked 3279 steps so far. How many more steps does he need to walk?

What Do You Think Now?

Look back at **What Do You Think?** on page 67. How have your answers and explanations changed?

Chapter Task

Counting Calories

Nutritionists recommend that you eat food containing 2200 to 2500 calories each day.

Joshua recorded everything he ate on Saturday and Sunday.

Joshua's Food Record

	Saturday			Sunday	
	Food	Calories		Food	Calories
Breakfast	restaurant breakfast	960		cereal with fruit	313
Lunch	10 chicken nuggets	510		tuna sandwich	361
	chocolate milkshake	850		vanilla yogurt	165
	french fries	610		apple	80
Snack	can of pop	150		hard-boiled egg	75
	cashew nuts	50		small juice box	95
Dinner	spaghetti	640		chicken and vegetables	847
	garlic bread	340		whole milk	150
	large pop	397		orange	65

(?) **What can you conclude about the food Joshua ate each day?**

A. Calculate the number of calories over or under the recommended number that Joshua ate each day.

B. Find data about calories in your favourite foods. Use the data to plan one day's food that contains 2200 to 2500 calories in total.

1. What numbers are missing from the addition table?

+	2	4	6	8	10	12
10	12	14	16		20	
20	22			28	30	32

A. 18, 24, 24, 26
B. 18, 20, 23, 24
C. 18, 20, 22, 24
D. 18, 22, 24, 26

2. How many squares are in shape 10 of this pattern?

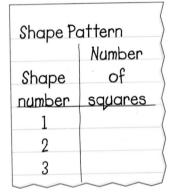

Shape Pattern	
Shape number	Number of squares
1	
2	
3	

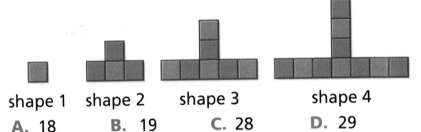

shape 1   shape 2   shape 3   shape 4

A. 18   B. 19   C. 28   D. 29

3. Every 5th clown in a parade of 75 clowns wears a rainbow wig. Every 4th clown has suspenders. How many clowns have a rainbow wig and suspenders?

A. 2   B. 3   C. 4   D. 5

4. What number does ▨ represent in this equation?

$50 = 48 +$ ▨

A. 50   B. 2   C. 52   D. 3

5. What number does ▨ represent in this equation?

▨ $- 34 = 43$

A. 77   B. 68   C. 9   D. 88

6. I have 6 thousands, 2 hundreds, and 3 tens. What number am I?

A. 6023   B. 6230   C. 623   D. 6023

7. What is one thousand eight hundred eighteen in standard form?
   A. 1888   B. 1818   C. 1880   D. 1808

8. What is 3504 in words?
   A. three thousand fifty-four
   B. three thousand five hundred four
   C. three hundred fifty-four
   D. thirty-five and four

9. Which number is greater than 2017?
   A. 2071   B. 218   C. 2007   D. 1071

10. Which number is between 6000 and 8000?
    A. 5999   B. 777   C. 8888   D. 7777

11. What is the difference between 4752 and 3000?
    A. 1752   B. 7752   C. 5052   D. 4452

12. Which number is the sum of 500 and 2800?
    A. 2300   B. 3300   C. 7800   D. 2850

13. Which numbers make this calculation true?
    A. 514 + 212
    B. 823 + 294
    C. 378 + 252
    D. 465 + 161

$$\begin{array}{r} \blacksquare\blacksquare\blacksquare \\ + \ \blacksquare\blacksquare\blacksquare \\ \hline 626 \end{array}$$

14. Yan bought 1000 seed beads. She used 312 to make bracelets. How many beads were left?
    A. 788   B. 712   C. 698   D. 688

15. In which case can an estimate be used?
    A. Josh tries to decide whether or not he has enough money to buy groceries.
    B. The clerk enters the amount of each item into the cash register.
    C. The clerk tells Josh how much he owes.
    D. Josh counts his change.

Canada's Medals at the Torino 2006 Olympic Winter Games

Number of medals

11
10
9
8
7
6
5
4
3
2
1
0

gold   silver   bronze

Medals

Canada's Medals at the Torino 2006 Paralympic Winter Games

gold

silver

bronze

Each ⬤ means 1 medal.

Data Relationships

GOALS

You will be able to

- interpret, compare, and construct pictographs and bar graphs

- find examples of graphs in the media

- use Venn diagrams and Carroll diagrams to sort and classify numbers

- use diagrams to solve problems

What do the 2 graphs tell you about the number of medals won by Canadian athletes at the 2006 Winter Games?

Getting Started

Sorting Creatures

Michael sorted creature cards from a game.
He used the **attribute** of *Ears* to sort the creatures into
2 groups. He made a **pictograph** and a **bar graph** to
show how he sorted.

Michael's Pictograph

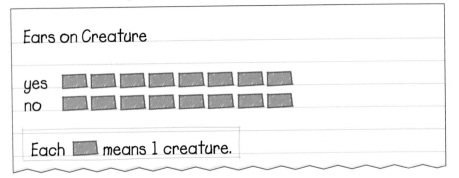

Ears on Creature

yes

no

Each �rectangle means 1 creature.

Michael's Bar Graph

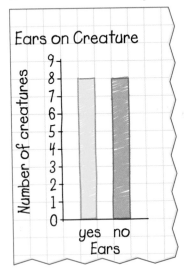

Ears on Creature

Number of creatures

9 8 7 6 5 4 3 2 1 0

yes no
Ears

 **What can a graph of sorted
creature cards show you?**

A. Do both of Michael's graphs show the same **data**? Explain how you know.

B. Choose another attribute and sort the creatures into 2 groups. Make a **tally chart** to show the number of creatures in each group.

C. Use the data in your tally chart to make either a pictograph or a bar graph. Draw the graph on grid paper. Make sure you use a **title**.

D. Describe 2 things your graph shows.

What Do You Think?

Do you *agree* or *disagree* with each statement? Explain your thinking.

1. This bar graph shows that more people like chocolate chips or berries on their ice cream than bananas or sprinkles.

2. You can use this graph to calculate the number of people who were asked about their favourite ice-cream topping.

3. The bars on this graph could have been drawn left to right (↔) instead of up and down (↕).

4. You can use this bar graph to help you make a pictograph.

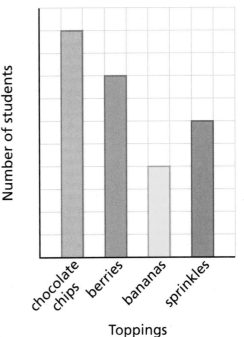

Favourite Ice-Cream Topping

Number of students

Toppings: chocolate chips, berries, bananas, sprinkles

Interpreting and Comparing Pictographs

GOAL

Interpret and compare pictographs that use different scales.

scale

The number represented by each unit or shape in a graph. For example, this is a scale for a pictograph.

Each ▮ means 4 students.

Hailey's class adopted a rattlesnake. They asked all the Grade 4 students to help them name the rattlesnake. Each student voted for 1 of 4 names. Hailey and Ken made pictographs of the vote using different **scales**.

CERTIFICATE OF ADOPTION
Adoption is hereby granted to
Hailey's Class
for ensuring the natural populations of Northern Pacific Rattlesnakes remain healthy by contributing to Nk'Mip's
Adopt a Snake Conservation Program.
HABITAT LEVEL

NK'MIP
Desert
Cultural
Centre

Charlotte Sanders, General Manager

Kyle Horner, Biologist

February 1, 2006
Date

Hailey's Pictograph

Rattlesnake Names

Sam Serpent ⊘⊘⊘⊘⊘⊘⊘ ⊘⊘⊘
Diamond Dan ⊘⊘⊘⊘⊘⊘⊘⊘⊘⊘⊘⊘⊘⊘⊘⊘⊘⊘⊘⊘⊘
Ray Rattler ⊘⊘⊘⊘⊘⊘⊘⊘⊘⊘⊘⊘⊘⊘⊘⊘⊘
Frank Fang ⊘⊘⊘⊘⊘

Each ⊘ means 1 vote.

Ken's Pictograph

Rattlesnake Names

Sam Serpent ○ ○ ○ ○ ○
Diamond Dan ○ ○ ○ ○ ○ ○ ○ ○ ○ ○
Ray Rattler ○ ○ ○ ○ ○ ○ ◐
Frank Fang ○ ○ ◐

Each ○ means 2 votes.

❓ **Do the graphs show the same data?**

A. How can you use the graphs to figure out the most popular name?

B. What is the scale of each graph?

C. How does skip counting by 2s help you interpret a pictograph that uses a scale of 2? Use Ken's pictograph to help you explain.

D. How many students voted? Explain.

E. Do the graphs show the same data? Explain.

Reflecting

F. Why does Ken's pictograph contain both whole and half circles?

G. Why does Ken's pictograph have half as many whole circles as Hailey's?

Checking

1. Diane and Annie asked this question:
   "Do rattlesnakes add a new part to their rattle each time they shed their skin?"
   They made pictographs of the data.

Diane's Rattlesnake Survey

Yes ◯

No ◯ ◖

Don't know ◯ ◯ ◯ ◖

Each ◯ means 10 votes.

Annie's Rattlesnake Survey

Yes ◯ ◯

No ◯ ◯ ◯

Don't know ◯ ◯ ◯ ◯ ◯ ◯ ◯

Each ◯ means 5 votes.

a) How are the 2 graphs the same? How are they different?
b) Do the graphs show the same data? Explain.
c) Use each graph to calculate the number of students who answered the question.
d) Why do you think neither student chose a scale of 1 ◯ means 1 vote?

Practising

2. The pictograph shows the number of male and female rattlesnakes studied one summer at the Nk'Mip Desert and Heritage Centre.

Rattlesnakes Studied

Each 🐍 means 5 snakes.

a) How many rattlesnakes were studied? Explain.
b) Does the graph show that more than twice as many males were studied as females? Explain.

3. Alaya's pictographs show the number of times her 4 friends each blinked in one minute. Do the pictographs show the same data? Explain how you know.

Blinks in One Minute

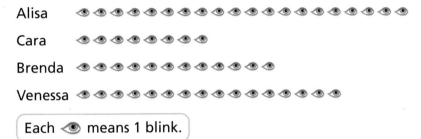

Each 👁 means 1 blink.

Blinks in One Minute

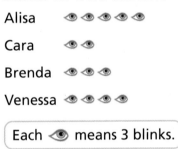

Each 👁 means 3 blinks.

4. Olivia's pictograph shows data about students in her school, but doesn't show the scale.

Students in Our School

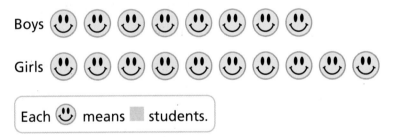

Each 😊 means ▨ students.

a) If Olivia's school has 4 more girls than boys, what scale did she use to make the graph? Explain.

b) If Olivia's school has 20 more girls than boys, what scale did she use to make the graph? Explain.

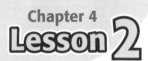

Constructing Pictographs

GOAL

Construct and interpret pictographs.

Alec's school visited the Royal Tyrrell Museum of Paleontology. Alec asked students which dinosaur they liked best.

Which Dinosaur Do You Like Best?	
Dinosaur	Number of students
Tyrannosaurus rex	60
Albertosaur	45
triceratops	25
dromaeosaurus	50
other	20

? **How can you display the data in pictographs with different scales?**

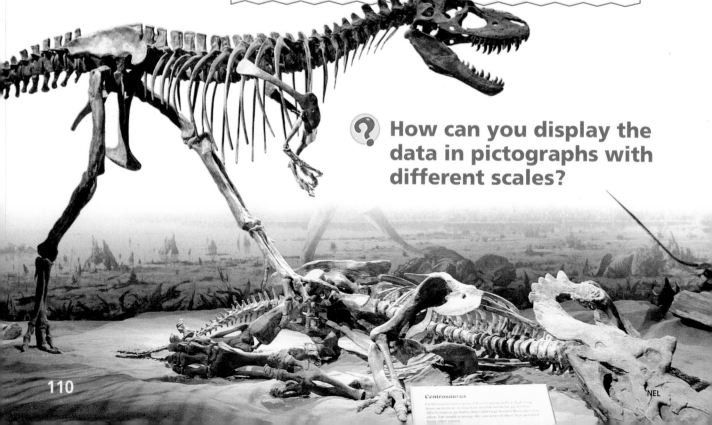

Alec's Pictograph

I chose a scale of 1 footprint means 10 students. It's easy to skip count by 10s, and I won't have to draw a lot of footprints.

I chose a footprint shape because it reminds me of dinosaurs. It's also easy to draw half footprints.

Which Dinosaur Do You Like Best?

Tyrannosaurus rex
Albertosaur

A. Explain how Alec might have used skip counting by 10s to figure out the number of whole and half footprints to draw for Albertosaur.

B. Copy and complete Alec's pictograph for the dinosaur data. Make sure you include the scale.

C. Describe 2 things the graph shows.

D. Make another pictograph of the data using a different scale and a different shape. Why did you choose that scale and shape?

Reflecting

E. How did you figure out the number of footprints to draw for triceratops?

F. What scales could you use so that your graph uses only whole shapes? Explain.

Checking

Visitors to the Dinosaur Show

Day	Number of visitors
Monday	150
Tuesday	200
Wednesday	200
Thursday	225
Friday	450

1. The chart shows the number of people who visited the dinosaur show from Monday to Friday.
   a) Make a pictograph of the data using a scale of 1 shape means 25 visitors.
   b) How did you figure out the number of shapes to draw for the number of visitors on Thursday?
   c) Describe 2 things your graph shows.
   d) How would your pictograph change if you used a scale of 1 shape means 10 visitors?

Practising

2. The pictograph shows the number of dinosaur bones found by different teams of scientists.

Number of Dinosaur Bones Found

Team A ○ ○ ○ ○ ○

Team B ○ ○ ○

Team C ○ ○

Team D ○ ○ ○ ○

Each ○ means 5 bones.

Team	Number of bones
A	
B	
C	
D	

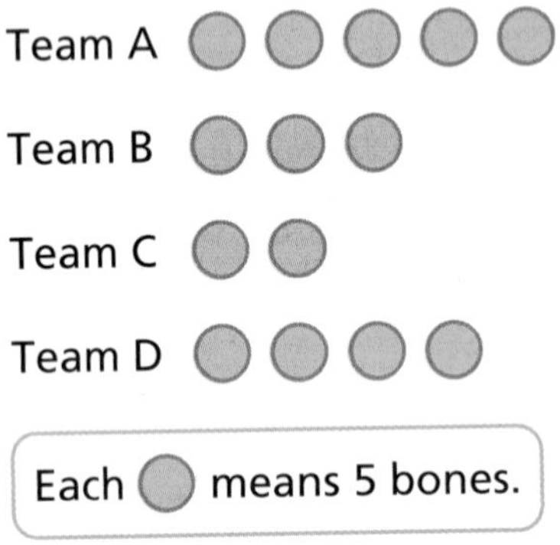

a) Which team found the greatest number of dinosaur bones? Which team found the least number of bones?
b) Use data from the pictograph to make a chart of the number of bones found by each team.
c) Use the chart to make a new pictograph. Choose a different shape and a different scale. Explain your choices.

3. Kalei asked 50 students whether they used a computer yesterday.

Did You Use a Computer Yesterday?

a) What is the scale for the graph? How do you know?

b) How many students used a computer yesterday?

c) Create a new pictograph using a different scale. Why did you choose that scale?

4. Shane wants to make a pictograph showing the number of cats and dogs available for adoption at an animal shelter.

a) Would you use a scale of 1 to display the data? Explain why or why not.

b) Would you use a scale of 20 to display the data? Explain.

c) Make a pictograph of the data. Explain your choice of shape and scale.

Animals for Adoption

Animal	Number of animals
dog	30
cat	45

5. Suppose you were experimenting with this spinner. In experiment A, you did 20 spins. In experiment B, you did 400 spins. You're making pictographs to show the number of times you landed on each colour.

a) For which experiment would you use a scale of 1? Explain.

b) For which experiment would you use a scale of 25? Explain.

6. How do you choose your scale when you construct a pictograph? Use the data in the table to help you explain.

Did You Watch the Hockey Game?

Yes	45
No	60

Interpreting and Comparing Bar Graphs

GOAL

Interpret and compare bar graphs that use different scales.

Diane's class has been collecting quarters, loonies, and toonies for the Variety Coin Drive.

Diane's Bar Graphs

I made 2 bar graphs about coins collected.

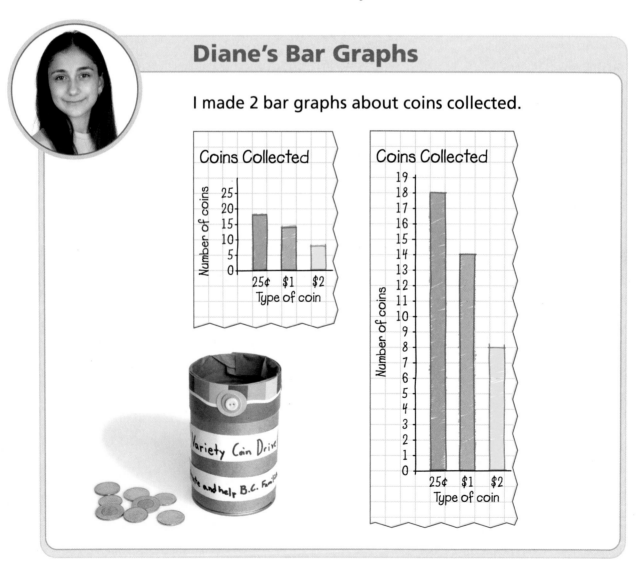

 **Do Diane's bar graphs show the same data?**

A. On each graph, which type of coin was donated most often? Explain.

B. What scale was used for each bar graph? Explain.

C. Use each graph to estimate the number of toonies collected.

D. Do both graphs seem to show the same data? Explain how you know.

Reflecting

E. Why is each bar on one graph shorter than the matching bar on the other graph?

F. If Diane used a scale of 100, would her graph be easy to read? Explain.

Ken's Graph

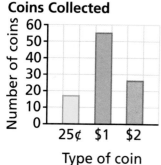

Desmond's Graph

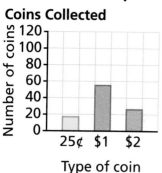

Checking

1. Ken and Desmond counted the coins in one collection can. The bar graphs show the results.
   a) How are the 2 graphs the same? How are they different?
   b) Can you use either graph to determine the exact number of loonies collected? Explain.
   c) Describe 2 things each graph shows you.
   d) Would you use a scale of 1 to make a bar graph of the data? Explain.

Practising

Reading Strategy

Look at the graphs. What inferences can you make?

2. Renée and Amelia each asked Grade 4 students in 2 schools this question: "Of the books you read this year, which was your favourite?" They each made a bar graph of the results.

Renée's Graph
Votes for Favourite Books

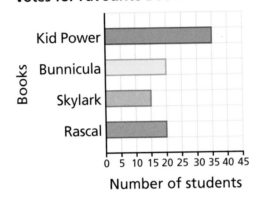

Amelia's Graph
Votes for Favourite Books

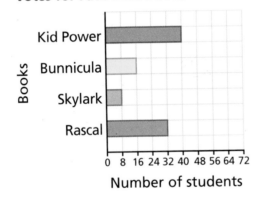

a) What was the most popular book for Grade 4 students? Explain how you know.

b) What is the scale of each graph? Explain.

c) Do the 2 graphs seem to show the same data? Explain how you know.

d) Use Renée's graph to estimate how many students answered her question. Show your work.

3. Wind turbines are used to produce electricity. Ryan's and Keifer's graphs show the number of wind turbines in Canada in 2005.

Keifer's Graph

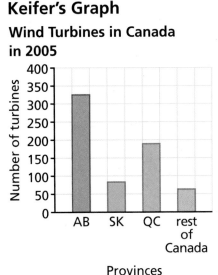

Ryan's Graph

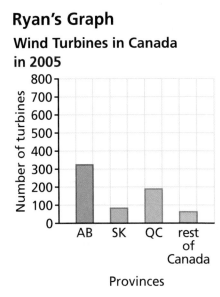

a) Use Keifer's graph to estimate how many more wind turbines are in Alberta than Saskatchewan.

b) Use Ryan's graph to estimate how many wind turbines there are in Canada.

c) Do the 2 graphs seem to show the same data? Explain how you know.

d) Why are the bars on Keifer's graph twice the height of the matching bars on Ryan's graph?

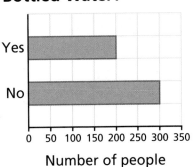

4. a) What is the scale of the bar graph about bottled water? Explain how you know.

b) How would you change the bar graph to a pictograph?

c) What is the same about bar graphs and pictographs? What is different?

Lesson 4

Constructing Bar Graphs

You will need
- a spinner
- a paper clip
- grid paper

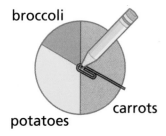

broccoli

potatoes   carrots

Vegetable Weather

Weather	Number of days
carrots	80
potatoes	73
broccoli	47

GOAL

Construct and interpret bar graphs.

The students in Ken's class read the book *Cloudy With a Chance of Meatballs* by Judi Barrett. The story is about a tiny town called ChewandSwallow where it never rains rain and it never snows snow. Instead, it rains and snows different types of food.

The class used a spinner to model some vegetable weather that might happen in ChewandSwallow.

The chart shows the results for 200 spins.

 How can you create and interpret a bar graph?

Ken's Bar Graph

I'll use a scale of 10 because the bars will fit on my graph paper.
I'll use **vertical** bars.
I'll use skip counting to label the numbers beside each line on the **vertical axis**.

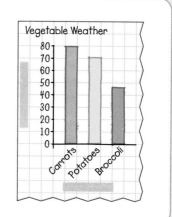

118

A. Explain how Ken might have used the numbers on the vertical axis to estimate the height of each bar.

B. What label should Ken add to each axis?

C. Conduct your own weather experiment. Spin the spinner 10 times to model 10 days of weather. Make a tally chart to show the number of times each type of weather happened.

D. Make a chart with the combined results for your whole class. Show the results in a bar graph.

E. On your graph, what is the most common type of vegetable weather?

Reflecting

F. What scale did you use for your graph? Why?

G. How did you use skip counting to write the numbers on the vertical axis?

Drink Weather

Weather	Number of days
milk	122
juice	78

Checking

1. Jon's class used this spinner to model drink weather in ChewandSwallow. The results are shown in the chart.
   a) Would you use a scale of 1 for the data? Explain.
   b) Make a bar graph to display the data. Use vertical or **horizontal** bars.
   c) What scale did you use? Explain your choice.

Practising

Reading Strategy

Read the question. Write any words you don't understand. Use the Glossary or a dictionary to look up the words.

2. The chart shows the attendance for 4 shows at a children's festival.

Shows at Children's Festival

Show	Attendance
Stumble On To	115
Cirkus Inferno	67
Anne Glover's Story String World	170
Music With Marnie	128

a) Make a bar graph to display the data. Use vertical or horizontal bars.
b) What scale did you use? Explain your choice.
c) About twice as many people saw Music With Marnie as Cirkus Inferno. Is it easier to use the graph or the chart to see this? Explain.

Pond Depth

Month	Depth (cm)
May	125
June	60
July	32
August	15

3. Barrett measured the level of a pond for a science fair experiment. The chart shows the depth of one spot in the pond over several months.
He drew a horizontal bar on grid paper to show the depth of the pond in May.

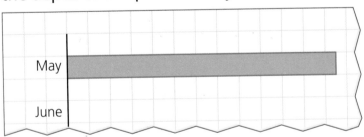

a) What scale did he use? How do you know?
b) Copy and complete Barrett's bar graph.
c) What does your bar graph tell you about the depth of the pond over the months Barrett measured?

Matching Data

Number of players: 2

How to play: Match 2 cards that show the same data.

• **Step 1** Shuffle the cards. Place them face down in an array of 6 rows of 4.

• **Step 2** Player 1 turns over 2 cards. If they show the same data, Player 1 keeps the cards and turns over 2 more cards. If the cards don't show the same data, Player 1 turns the cards back over.

• **Step 3** Take turns until all the cards have been taken.

The player with the most cards is the winner.

Diane's Turn

I turned over 2 cards. They both show the same data. So I keep these 2 cards and can now turn over 2 more cards.

Yes	30
No	45

Yes ◆◆◆
No ◆◆◆◆◀

Each ◆ means 10.

Mid-Chapter Review

Frequently Asked Questions

Q: How can you tell if graphs show the same data?

A1: Use the scale on each pictograph to calculate the number in each category. For example, both graphs show 11 Yes and 5 No, so the data are the same.

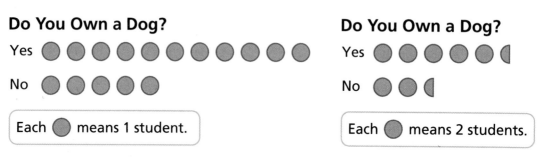

Do You Own a Dog?

Yes ⬤⬤⬤⬤⬤⬤⬤⬤⬤⬤⬤

No ⬤⬤⬤⬤⬤

Each ⬤ means 1 student.

Do You Own a Dog?

Yes ⬤⬤⬤⬤⬤◖

No ⬤⬤◖

Each ⬤ means 2 students.

A2: Use the scale on each bar graph to calculate or estimate the number in each category. For example, both graphs show between 25 and 30 heads and between 20 and 25 tails, so the data seem to be the same.

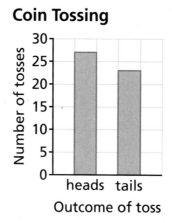

Coin Tossing

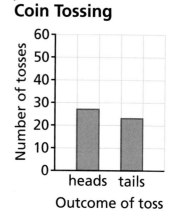

Coin Tossing

Q: How can you construct graphs using data in a chart?

A1: To make a pictograph, begin by selecting a shape and a scale.

For example, to show the hockey data, you might choose a scale where each hockey puck shows 1, 2, or 5 students because you know how to skip count by these numbers.

If you use a scale of 1 puck means 5 students, you need only 8 whole pucks to show all the data. The rows are easy to read and compare.

Did You Watch the Hockey Game Last Night?

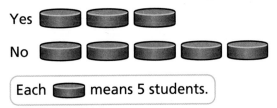

Did You Watch the Hockey Game Last Night?

Yes	卌 卌 卌
No	卌 卌 卌 卌 卌

A2: To make a bar graph, begin by selecting a scale. For example, to show the die-rolling data,

- the bars would be too tall with a scale of 1 or 2.
- the bars would be too short with a scale of 10.
- the bars would be easy to see and compare with a scale of 5.

Once you select a scale, such as 5, you skip count and write these numbers at each line on an axis. Then, write the categories "even number" and "odd number" on the other axis. Next, use the numbers along the axis to determine the heights of the bars.

- The top of the bar for even numbers is along a line between 25 and 30.
- The top of the bar for odd numbers is along a line between 30 and 35.

Finally, label each axis and give the graph a title.

Die Rolling

Even number	28
Odd number	32

Die Rolling

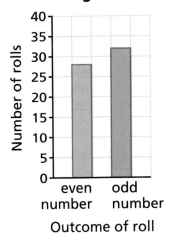

Practice

Lesson 1

1. The 2 pictographs show the number of aluminum cans collected for recycling by some Grade 4 students in a week.

Cans Collected

Tim

Rosie

Aputik

Each 🛢 means 10 cans.

Cans Collected

Tim

Rosie

Aputik

Each 🛢 means 1 can.

a) Why does one pictograph show only whole can pictures?
b) Why does the other pictograph show both whole and half cans?
c) Do the graphs show the same data? Explain how you know.
d) Describe 2 things one of the graphs shows you.

Lesson 2

What Students Are Reading

Books 🗋🗋🗋🗋🗋🗋

Newspapers 🗋🗋

Magazines 🗋🗋🗋

Comics 🗋🗋🗋🗋🗋

Each 🗋 means 10 students.

2. The pictograph shows data about the types of reading material students chose in the library.
a) Use the pictograph to create a chart of the data.
b) Use the data in the chart to draw another pictograph with a different scale and a different shape.
c) Explain why you chose this scale and shape.
d) Use one of the pictographs to calculate the total number of students. Show your work.

3. Each graph shows the most popular baby boy names in one city in 2005.

Most Popular Baby Boy Names

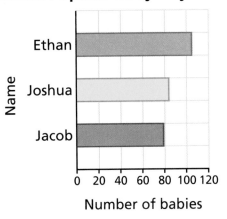

Most Popular Baby Boy Names

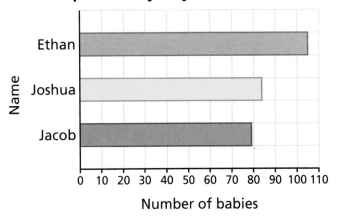

a) Which name is most popular? How do you know?

b) Use both graphs to estimate the number of babies named Jacob. Show your work.

c) Do both graphs seem to show the same data? Explain.

Lesson 4

4. A sandwich shop kept track of sales for a week.

Number of Sandwiches Sold

Sandwich	turkey	vegetarian	roast beef	tuna	meatball
Number sold	68	37	85	105	96

a) Would you use a scale of 1 to make a bar graph to show the data? Explain.

b) Make a horizontal or vertical bar graph to show the data. Remember to include the title and labels for the axes.

c) What scale did you use? Explain your choice.

d) Describe 2 things your graph shows.

Graphs in the Media

GOAL

Find and interpret graphs used in the media.

Alec completed an online weekly survey at the official website of the 2010 Winter Olympic Games in Vancouver. This survey asked people to name their favourite Paralympic event. When he completed the survey, a bar graph was displayed.

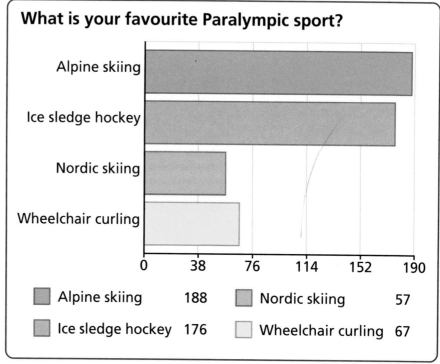

What is your favourite Paralympic sport?

Alpine skiing	188
Ice sledge hockey	176
Nordic skiing	57
Wheelchair curling	67

What questions can you answer about graphs you find in newspapers, magazines, or on the Internet?

Logic Puzzles

Lewis Carroll, who wrote *Alice in Wonderland,* created many logic puzzles and ways to use diagrams to solve puzzles.

You will need
- grid paper

1. Copy and complete the chart to solve this puzzle.

   Jack drives at least 1 of his grandchildren to soccer, hockey, and piano each week. Each grandchild does a different activity.

   Hobbies

	Barrett	Maddy	Emily
Soccer	✔		
Hockey	X		
Piano	X		

   - Maddy doesn't play any sports.
   - Barrett plays a sport that uses a ball.

   If Emily does only 1 activity, what is it?

2. Peter, Shane, and Kent have a horse named Dusty. The horse is ridden once each day of the week except on Sunday.
   - Peter rides every other day starting on Monday.
   - Shane rides on Tuesday and Thursday.
   - What day(s) does Kent ride?

Riding Days

	Sun	Mon	Tues	Wed	Thurs	Fri	Sat
Peter							
Shane							
Kent							

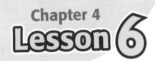

Lesson 6 Using Venn Diagrams

Using Venn Diagrams

You will need

- Venn Diagrams
  (blackline masters)

GOAL

Sort and classify numbers using Venn diagrams.

Venn diagram

A diagram that uses shapes such as circles to show relationships

Michael, Hailey, and Diane attended a math fair. Each student can win a prize by correctly sorting 12 numbers using different kinds of **Venn diagrams**.

 How can you sort the numbers using Venn diagrams?

Michael's Sorting

I'll sort the numbers using this Venn diagram with 2 separate circles.

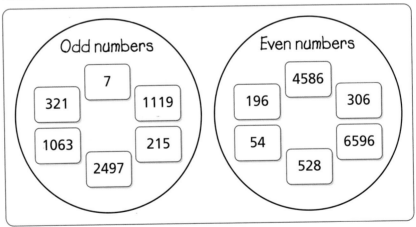

Numbers

Odd numbers: 7, 321, 1119, 1063, 215, 2497

Even numbers: 4586, 196, 306, 54, 6596, 528

Hailey's Sorting

I'll sort the numbers using this Venn diagram with overlapping circles.

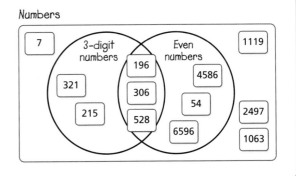

Diane's Sorting

I'll sort the numbers using this Venn diagram with a small circle inside an oval.

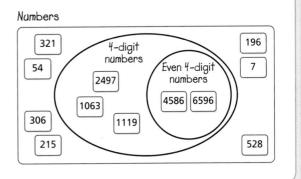

Reflecting

A. Choose one number from the Venn diagrams. Explain how each student placed this number in each Venn diagram.

B. Why was Michael able to do his sorting using separate circles instead of overlapping circles?

C. In Diane's Venn diagram, why is the small circle inside the oval instead of outside the oval?

Checking

1. Stacy was given a Venn diagram and some numbers to sort. Copy and complete Stacy's sorting.

Numbers

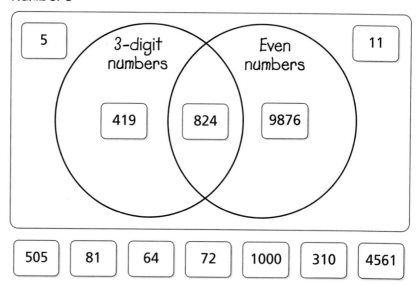

| 505 | 81 | 64 | 72 | 1000 | 310 | 4561 |

Practising

Numbers 1 to 20

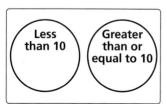

Numbers 1 to 20

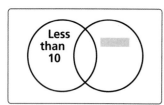

2. Janet used a Venn diagram with separate circles to sort the numbers from 1 to 20.
   a) Sort the numbers using a Venn diagram like Janet's.
   b) Are there any numbers outside the circles? Explain.
   c) Janet wants to sort the numbers again using a Venn diagram with overlapping circles. What label can she use so some numbers will be in the overlap?
   d) Use a Venn diagram with overlapping circles to sort the numbers.

3. Corina sorted the numbers from 1 to 20 using this Venn diagram. She created a puzzle by removing the labels from each circle.

Numbers 1 to 20

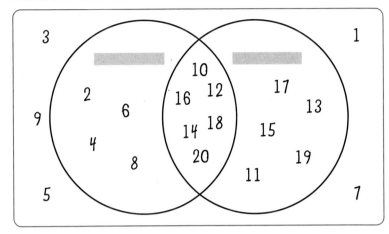

a) What labels might be on each circle? Explain.
b) Use a Venn diagram to sort the numbers from 1 to 20 using different labels. Create a puzzle for another student to solve by removing the labels.

4. Emily listed the number of students in each classroom in her school: 15, 19, 20, 25, 26, 28, 31, 35. Sort these numbers, or numbers from your school, using all 3 types of Venn diagrams. Include labels.

5. Barrett wants to put these pairs of labels on 2 different Venn diagrams to sort some whole numbers.

a) Which 2 labels should he put on the Venn diagram with overlapping circles? Explain.
b) Which 2 labels should he put on the Venn diagram with separate circles? Explain.

Lesson 7 Using Carroll Diagrams

You will need
- grid paper

GOAL

Sort and classify numbers using Carroll diagrams.

Carroll diagram

A chart with rows and columns that shows relationships.

Annie is sorting the addresses of 12 students in her class.

291   515   1610   4586   69   131
1883   17   9   44   162   2222

 How can you use a Carroll diagram to sort the addresses?

Annie's Carroll Diagram

The labels on the diagram tell the **attributes** for sorting the addresses.

Addresses

	Fewer than 4 digits	4 digits or more
Odd	291, 515	
Not odd		1610

A. Copy Annie's Carroll diagram and finish sorting the addresses. Explain how you placed each number.

B. Choose 2 more numbers for each part of the Carroll diagram. Add them to your diagram.

C. Sort the address numbers of at least 12 students in your class in a Carroll diagram using Annie's labels.

Reflecting

D. Why can't you put the same address in more than one row or more than one column?

E. Are there any addresses that can't be placed in Annie's Carroll diagram? Explain.

Checking

1. a) Sort the addresses you used in Part C using a Carroll diagram like the one to the left.

b) Are there any addresses that can't be placed in this Carroll diagram? Explain.

Practising

2. a) Sort these numbers using a Carroll diagram like the one to the left.

25, 37, 15, 24, 17, 200, 36, 99, 168, 320, 10, 1280

b) Choose and write 2 more numbers in each cell of the Carroll diagram.

3. a) Emily sorted numbers using a Carroll diagram like the one to the left. What labels are missing?

b) Choose and write 2 more numbers in each cell of the Carroll diagram.

4. Use a Carroll diagram to sort the numbers in this Venn diagram using the same attributes.

Addresses

	Fewer than 4 digits	4 digits or more
Has the digit 1		
Does not have the digit 1		

Numbers

	Tens digit is even	Tens digit is not even
3 digits or more		
Fewer than 3 digits		

Numbers

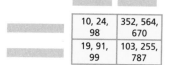

	10, 24, 98	352, 564, 670
	19, 91, 99	103, 255, 787

Numbers

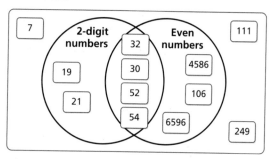

Chapter 4
Lesson 8

Solving Problems Using Diagrams

You will need
- Venn Diagrams (blackline masters)
- grid paper

Use diagrams to solve problems.

A scientist is studying 40 black bears. Some of them have white coats. 16 males have black coats. 17 bears are male. 36 bears have black coats.

 How many of the bears are female with white coats?

Alec's Solution

I'll use a Carroll diagram to solve the problem.
16 male bears have black coats.
17 bears are male.
17 − 16 = 1, so I know 1 male bear has a white coat.

Black Bears

	Black coats	White coats
Male	16	1
Female	20	3

36 bears have black coats.
36 − 16 = 20, so I know 20 of the bears with black coats are female.
There were 40 black bears in the study.
16 + 20 + 1 = 37, so 3 female bears must have white coats.

Reflecting

A. How can you use Alec's Carroll diagram to check his answer?

B. Could Alec begin by writing the total number of male bears or the total number of black bears in the Carroll diagram? Explain.

Bears

	Cub	Not a cub
Male		
Female		

Checking

1. A scientist is studying 30 bears. 7 cubs are male. 10 bears are male. 11 cubs are female. How many female bears are not cubs? Show your work.

Practising

Shapes

	Squares	Triangles
Red		
Blue		

2. 40 squares and triangles are on the floor. 8 squares are red. 8 triangles are blue. The total number of squares is 29. How many triangles are red? Show your work.

3. Solve Question 2 using a Venn diagram like this one. Show your work.

Shapes

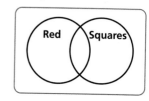

4. Ben has a collection of 33 hockey cards. 11 show Canadian players, and 18 show goalies. 7 are Canadian goalies. How many cards show players who are not Canadian and not goalies? Use a Venn diagram or a Carroll diagram to solve the problem.

Students

	Boys	Girls
Wear glasses		
Don't wear glasses		

5. a) Use data about the students in your class to complete a Carroll diagram like this.
   b) Erase one of the numbers. Use the rest of the Carroll diagram to create a problem. Ask another student to solve your problem.

Chapter Review

Frequently Asked Questions

Q: What are some ways to sort numbers in Venn diagrams?

A: You can use different Venn diagrams to sort the same numbers.

Numbers 1 to 10

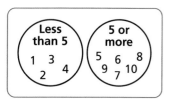

• For example, you can sort the numbers 1 to 10 into 2 separate groups: a group of numbers less than 5 and a group of numbers 5 or more. The circles are separate.

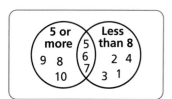

• You can sort the numbers into groups that overlap: some numbers are 5 or more, some are less than 8, and some are both. The circles overlap.

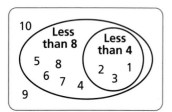

• You can sort the numbers into 3 groups: some numbers are less than 8, some are less than 4, and some are neither. One circle is inside an oval and some numbers are outside both the oval and circle.

Q: How can you use a Carroll diagram to solve a problem?

A: A Carroll diagram can be used to solve a problem when the information in the problem involves 2 attributes.

For example, there are 28 students in Natalie's class. 15 students have cats. 10 students have dogs. 7 students have both. How many students don't have a cat or a dog? Answer: **10**

Pets

	Cat	No cat
Dog	7	10−7
No dog	15−7	

Practice

Lesson 1

1. Each pictograph shows the number of greeting cards reused by Ava's class for craft projects.

Recycled Cards

Friendship 🔲🔲

Thank you 🔲🔲▮

Holiday   🔲🔲🔲

> Each 🔲 means 4 cards.

Recycled Cards

Friendship 🔲🔲🔲🔲🔲🔲🔲

Thank you 🔲🔲🔲🔲🔲🔲🔲🔲🔲

Holiday   🔲🔲🔲🔲🔲🔲🔲🔲🔲🔲🔲🔲

> Each 🔲 means 1 card.

    a) Use each graph to determine how many thank-you cards the class recycled.

    b) Do both graphs seem to show the same data? Explain.

Colour	Number
red	ℍℍℍℍ‖
blue	ℍℍℍℍ
green	ℍℍ‖
other	ℍℍ‖‖

Lesson 2

2. Emily's tally chart shows her friends' favourite colours.

    a) Use the data in the chart to make a pictograph.

    b) What scale did you use? Explain your choice.

Lesson 3

3. Each graph shows the number of events in the past 4 Winter Olympic Games.

Number of Olympic Events

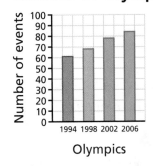

Number of Olympic Events

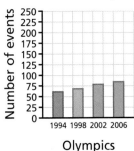

    a) Do the graphs seem to show the same data? Explain.

    b) Describe 2 things one of the graphs shows.

Growing Seeds	
Wednesday	18
Friday	61
Sunday	90
Tuesday	90

Lesson 4

4. On Monday, Justine placed 100 seeds on a damp paper towel. Every other day, she counted the total number of seeds that were growing. The chart shows her results.
   a) Make a bar graph to display the data.
   b) What scale did you use? Explain your choice.
   c) What does your graph show about the growth of the seeds?

Numbers 10 to 25

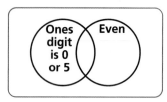

Lesson 6

5. a) Use a Venn diagram like the one at the left to sort the numbers from 10 to 25.
   b) Are there numbers outside the 2 circles? Explain.

Numbers

	2-digit numbers	3-digit numbers
Has the digit 5		
Has no digit 5		

Lesson 7

6. a) Sort these numbers using a Carroll diagram like the one at the left.
   249, 787, 15, 30, 675, 54, 19, 95, 564, 61, 353, 103
   b) Choose and write 2 more numbers in each part of the Carroll diagram.

Students

	Born in the Year of the Rat	Born in the Year of the Ox
Boys		
Girls		

Lesson 8

7. After reading the book *The Great Race* by David Bouchard, 48 students learned that each was born in the Year of the Rat or Ox. 25 students were boys. 7 boys were born in the Year of the Rat. 9 girls were born in the Year of the Rat. How many girls were born in the Year of the Ox?

What Do You Think Now?

Look back at **What Do You Think?** on page 105. How have your answers and explanations changed?

Chapter Task

Task Checklist
✔ Did you use math language?
✔ Did you show all your steps?
✔ Did you explain your thinking?

Promoting Fire Safety

After some firefighters visited her classroom, Annie decided to make a poster about fire safety.

She asked Grade 4 students in her school to answer this question: "How many fire extinguishers do you have in your home?"

She used the answers to make a bar graph.

Fire Extinguishers at Home

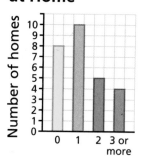

? **What question and graph would you include in a poster about fire safety?**

A. Use the graph to calculate the number of students in Annie's study.

B. Make a new bar graph or pictograph of Annie's data. Use a different scale. Explain your choice of scale.

C. Write a question about fire safety. Ask more than 30 students to answer your question. Show the data in a tally chart.

D. Make a pictograph or bar graph to show your data from Part C. Explain your choice of scale.

E. Describe 2 things your graph shows.

NEL

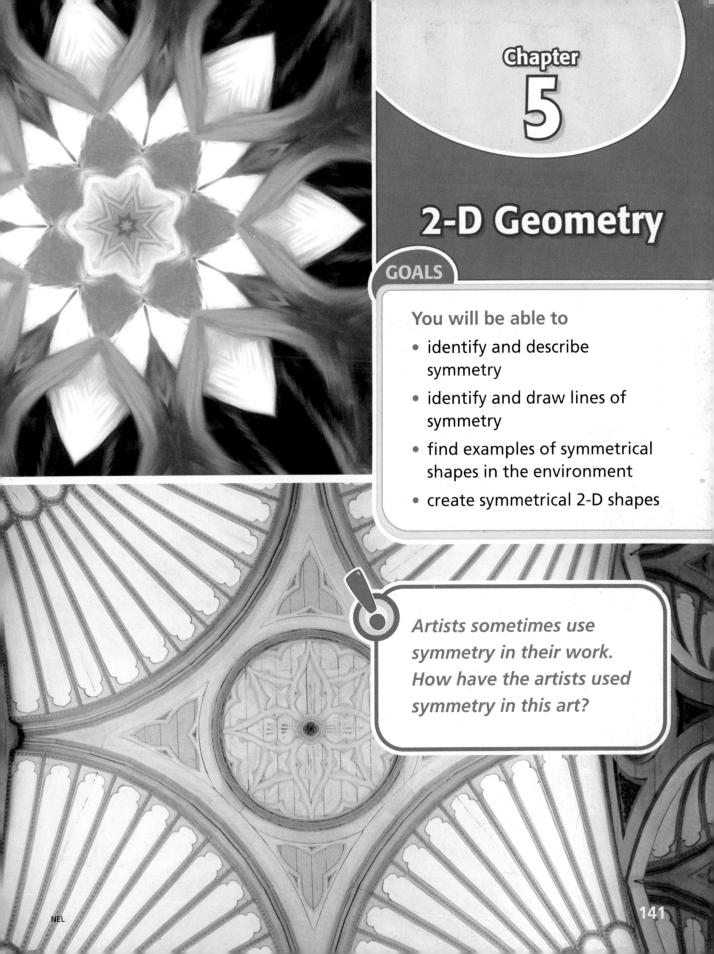

Chapter 5

2-D Geometry

GOALS

You will be able to

- identify and describe symmetry
- identify and draw lines of symmetry
- find examples of symmetrical shapes in the environment
- create symmetrical 2-D shapes

Artists sometimes use symmetry in their work. How have the artists used symmetry in this art?

Getting Started

- Traffic Signs (blackline master)
- scissors
- a ruler

Exploring Polygons

Emily cut out **polygons** like the shapes of these signs.

She cut one in half by accident.
Emily noticed that both halves have the same number of sides as the original shape.

? **Which shapes can be cut so that each half has the same number of sides as the original shape?**

A. Name all the traffic sign shapes by the number of sides they have.

B. Cut out polygons shaped like the traffic signs.

C. Imagine cutting each shape in half. You may have to imagine different cuts to get halves. Then use 1 straight cut to create the 2 halves.

D. How do you know each part is a half?

E. Name each new shape by the number of sides it has.

F. Which new shapes have the same number of sides as the original shapes?

What Do You Think?

Do you *agree* or *disagree* with each statement?
Explain your thinking.

1. Two halves of a shape always look alike.

2. When you fold a rectangle on a diagonal, the 2 parts match.

3. Some shapes can be folded in half in only one way.

Lines of Symmetry

GOAL

Identify lines of symmetry by folding.

Luis needs a square piece of paper to fold into a whale. Luis isn't sure that his piece of paper is square. He doesn't have a ruler, so he can't measure it to make sure.

 Is Luis's piece of paper a square?

Luis's Shape

I know a square has 4 equal sides.
I'll fold the paper to see if all the sides match.

The halves fit on top of each other.
They are the same size and shape.
So, each fold line is a **line of symmetry**.

A line that divides a
2-D shape in half so
that if you fold the
shape on this line,
the halves will
match

lines of
symmetry

This shape has
2 lines of symmetry.

A. Cut out a shape like Luis's.

B. Fold the shape in different ways to match the sides.

C. Mark the fold lines on the shape. How do you know these fold lines are lines of symmetry?

D. Is Luis's shape a square? Explain how you know.

Reflecting

E. Do you think Luis's shape has other lines of symmetry? Explain.

F. How can you locate a line of symmetry?

Checking

1. a) Cut out shapes like these. Which one does not have a line of symmetry?

b) Identify and draw the lines of symmetry on the other shapes by folding your cutouts.

Practising

2. a) Cut out triangles like these. Which ones have lines of symmetry?

b) Draw all the lines of symmetry in your triangles.

3. a) Cut out shapes like these. Identify lines of symmetry in each shape by folding.

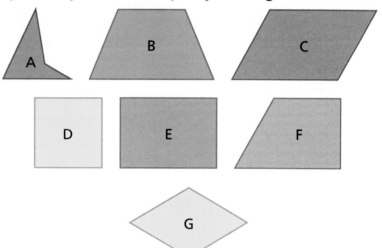

b) Draw all the lines of symmetry on the shapes.

4. Which line in this shape is a line of symmetry? How do you know?

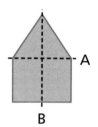

5. Which of these shapes probably do not have a line of symmetry? Explain your thinking.

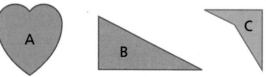

6. Which dotted line is not a line of symmetry for the square? Explain how you know.

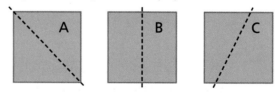

Using a Symmetry Tool

You will need
- a transparent mirror
- scissors
- pattern blocks
- a ruler
- dot paper
- Flags (blackline master)

GOAL

Use a transparent mirror to test for symmetry and complete a symmetrical shape.

symmetrical

A way of describing a 2-D shape with at least one line of symmetry.

lines of symmetry

Jade and her grandfather went to a Ukrainian festival. They noticed **symmetrical** patterns there.

Jade placed a transparent mirror on the line of symmetry of one of the shapes. The shape on one side matched the shape on the other.

Jade put the transparent mirror on a line that was not a line of symmetry, and the 2 sides didn't match.

Jade wants to create symmetrical shapes. She started with these shapes.

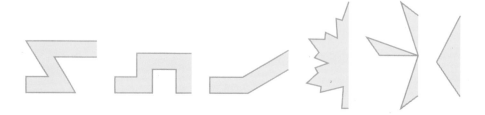

 How can you create symmetrical shapes?

Jade's Symmetrical Shapes

First, I'll visualize where the line of symmetry might be on my shape.

I'll put the transparent mirror on the line of symmetry.

I'll trace the outline that appears on the other side of the mirror.

I'll cut out the shape.

I'll check for symmetry by folding.

Reflecting

A. How can you tell if a transparent mirror is on a line of symmetry?

B. How can you create a symmetrical shape using a transparent mirror?

Checking

1. Which shapes are symmetrical? Use a transparent mirror to check.

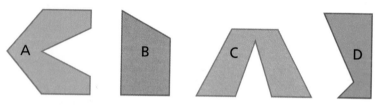

2. **a)** Choose one of Jade's shapes. Use it to create a symmetrical shape.
   b) Draw the line of symmetry.

Practising

3. Which shapes are symmetrical? Use a transparent mirror to check.

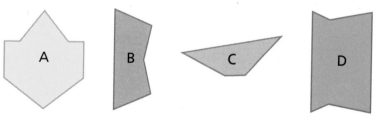

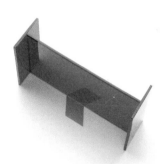

4. **a)** Cut out the shape of a pattern block. Create a new symmetrical shape by placing a transparent mirror along one side of the shape.
   b) Draw and cut out the new shape.
   c) Draw the line of symmetry on the new shape.
   d) Check for symmetry by folding.

Reading Strategy

Picture each shape in your mind. Can you see a line of symmetry?

5. Sketch shapes like these. Draw each line of symmetry. Use a transparent mirror.

a)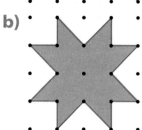

c)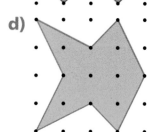

b)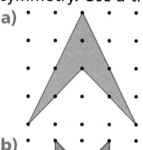

d)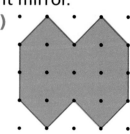

6. a) Use a transparent mirror to create a symmetrical 2-D shape.
   b) Draw the line of symmetry.
   c) Check for symmetry by folding.

7. a) Identify a line of symmetry in each flag. Use a transparent mirror.

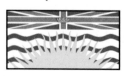  

   b) Check your work by cutting and folding paper copies of flags like these.

8. What different shapes can you make from a triangle like this by placing a transparent mirror on it?

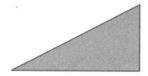

9. Suppose you use a transparent mirror to create a symmetrical shape. How do you know that the shape will have at least 2 equal sides?

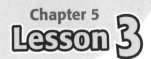
Identifying Symmetrical Shapes

GOAL

Identify and sort shapes in the environment using symmetry.

Ethan found this picture in the book *Imagine a Day* by Sarah L. Thomson. He can see symmetry in this picture.

? **Where can you find symmetry in the faces of 2-D shapes in your environment?**

Mid-Chapter Review

Frequently Asked Questions

Q: **How can you identify a line of symmetry of a shape?**

A1: You can fold the shape to see whether there is a fold line that will divide the shape into matching halves.

A2: You can place a transparent mirror on a line on the shape to see whether one side fits exactly on top of the image on the other side.

Practice

Lesson 1

1. Cut out shapes like these. Identify and draw the lines of symmetry by folding.

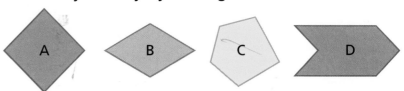

Lesson 2

2. a) Which letters have a line of symmetry?
   b) Write the letters and draw their lines of symmetry.

A B D F M
O P R T V

3. Complete shapes like these to make symmetrical 2-D shapes. Use a transparent mirror. Make the dotted lines the lines of symmetry.

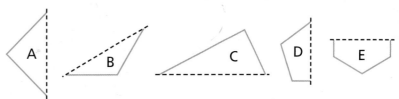

Lesson 3

4. a) Find a symmetrical 2-D shape in the classroom.
   b) Sketch the shape.
   c) Draw the lines of symmetry on it.

Chapter 5

Lesson 4

Counting Lines of Symmetry

You will need
- a ruler
- a transparent mirror
- scissors
- Kite Shapes
  (blackline master)
- 2-D Shapes 6
  (blackline master)

GOAL

Sort shapes according to the number of lines of symmetry.

During *La fête du Solstice* in Vancouver, Cory wants to buy a kite. Cherise says a kite will fly better if it has lines of symmetry.

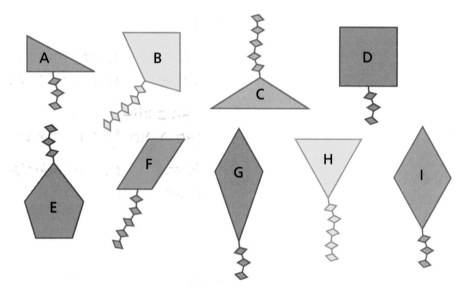

? Which kite should Cory buy?

Cory's Reasoning

I'll figure out how many lines of symmetry each kite has. Then I'll pick kites with more lines of symmetry.

154

NEL

A. How many lines of symmetry do you think each kite has?

B. Test your predictions using a transparent mirror. Were your predictions correct or incorrect? Explain.

C. Complete a chart like this one.

Kite	Number of lines of symmetry	Number of equal sides
A	0	

D. Which kite should Cory buy? Why?

Reflecting

E. Look at columns 2 and 3 for the kites in the chart. What do you notice?

F. Can a kite with 0 equal sides have a line of symmetry? Explain your thinking.

G. How can you tell by looking whether a shape might have more than 1 line of symmetry?

Checking

1. **a)** Predict whether each kite has 0, 1, or more than 1 line of symmetry. Explain.
   b) Test your predictions. Cut out kite shapes like these. Draw the lines of symmetry.

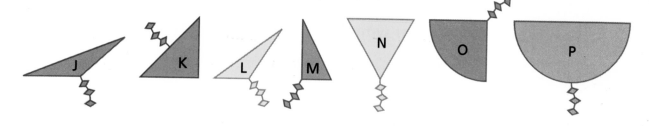

Practising

2. **a)** Predict whether each shape has 0, 1, or more than 1 line of symmetry.

   b) Test your predictions. Cut out shapes like these. Draw the lines of symmetry.

   c) Sort the shapes in a chart like this.

Shape	Number of lines of symmetry	Number of equal sides

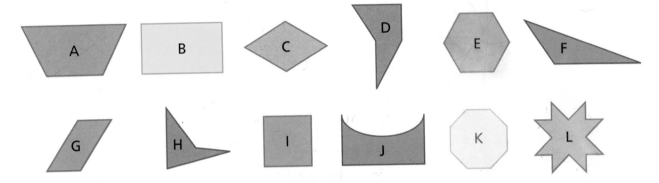

Reading Strategy

What questions can you ask to help you understand the problem?

3. One shape in each group is different from the others. Think about lines of symmetry. Which shape is different? Explain your thinking.

   a)

   b)

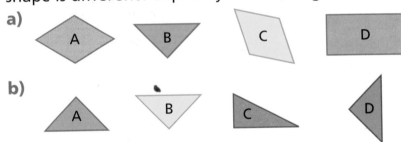

4. Do shapes with more sides usually have more lines of symmetry? Explain.

NEL

Tangram Shapes

You will need
- Tangram (blackline master)
- scissors
- a transparent mirror

Number of players: 2 or more

How to play: Use tans to make symmetrical shapes.

- **Step 1** Select one of the shapes shown below.

- **Step 2** Each player tries to form the shape using all 7 tans from the tangram.

- **Step 3** The first player to form the shape gets 1 point.

- **Step 4** Check that the shape is symmetrical by tracing it and drawing lines of symmetry.

Continue playing until all 4 shapes have been made.

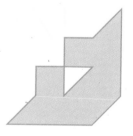  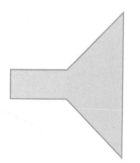 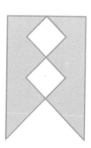

Communicating about Symmetry

You will need
- Ethan's Shapes (blackline master)
- a transparent mirror
- scissors

GOAL

Describe symmetry in 2-D shapes.

Ethan coloured these shapes for a poster. He is writing descriptions for a display on parent night.

A B C D E

? **How can you describe the symmetry in Ethan's shapes?**

Ethan's Description

Shape A is a **regular polygon**. I can fold it in half, so it is symmetrical. When I put a transparent mirror on the fold line in shape A, the shape and colours match. So, the colours are symmetrical.

For shape B, the shape is symmetrical but the colours are not.

shape A

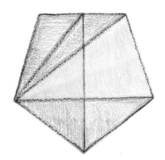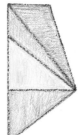

shape B

A. Use the Communication Checklist. What was good about Ethan's description?

B. How can you improve Ethan's description?

Reflecting

C. How does a diagram help to show what Ethan was thinking?

Checking

1. Cut out shapes like Ethan's. Describe the symmetry in each shape. Use the Communication Checklist.

Practising

2. **a)** Choose any 2 shapes. Describe any symmetry in them. Don't identify the shapes by name, by letter, or by colour.

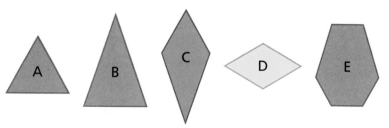

   b) Trade your descriptions with a partner. Are your descriptions clear enough that your partner can identify the shapes you described?
   c) Do your descriptions need improvement? Use the Communication Checklist.

3. **a)** Choose any shape. Describe any symmetry in the colours and the shape. Don't identify the shape by name or by letter.

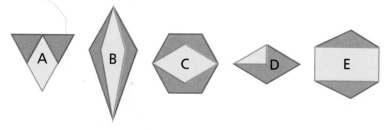

   b) Trade your description with a partner. Is your description clear enough that your partner can identify which shape you described?
   c) Does your description need improvement? Use the Communication Checklist.

Creating Symmetrical Shapes

You will need
- pattern blocks
- pattern-block grid paper

GOAL

Create symmetrical shapes.

Olivia is creating a shape using pattern blocks. Olivia thinks the shape will look better if it is symmetrical. She will use the pattern blocks first and then create the shape without the blocks.

? **What symmetrical shapes can you create?**

Folding Paper Shapes

This piece of paper was folded twice. There's a triangle piece missing from the corner. But if you open it up, the missing piece is a four-sided shape, not a triangle.

1. Draw what each piece of paper looked like before it was folded.

   a) This piece of paper was folded to the right once.

   b) This piece of paper was folded up once.

   c) This piece of paper was folded to the left once, and then folded up once.

   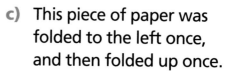

   d) This piece of paper was folded up once, and then folded to the right once.

Chapter Review

Frequently Asked Questions

Q: How might you predict whether a 2-D shape is symmetrical?

A: A 2-D shape might be symmetrical if it has at least 1 pair of equal sides. To be sure, you have to use a transparent mirror or fold to see if the halves match and if the sides of the halves match. For example, shapes A, B, and C are symmetrical and have at least 1 pair of equal sides.

A   **B**   **C**

 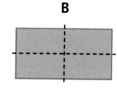

D

However, if a shape does have equal sides, you can't be sure that it is symmetrical. Shape D shown at the left has 2 pairs of equal sides, but it is not symmetrical.

Q: How many lines of symmetry can a shape have?

A: A shape can have 0, 1, or more than 1 line of symmetry. For example, Shape E has 0 lines of symmetry. Shape F has 1 line of symmetry. Shape G has more than 1 line of symmetry.

E   **F**   **G**

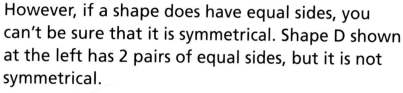

Practice

Lessons 1, 2, and 3

1. a) Identify a symmetrical 2-D shape in your classroom. Sketch the shape.
   b) Show that the shape is symmetrical.

2. Create a symmetrical shape. Start with a shape like this. Use one of the sides as the line of symmetry.

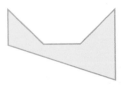

Lesson 4

3. a) Choose a pattern block. Predict whether the face of the block has 0, 1, or more than 1 line of symmetry.
   b) Test your prediction. Cut out a pattern block shape. Draw all the lines of symmetry.

Lesson 5

4. a) Explain how to find the lines of symmetry on the faces of all the pattern blocks in Question 3.
   b) Sort the pattern blocks according to the number of lines of symmetry on their faces.

Lesson 6

5. Use 2 or more pattern blocks.
   a) Create a shape with 0 lines of symmetry.
   b) Create a shape with 1 line of symmetry.
   c) Create a shape with more than 1 line of symmetry.

What Do You Think Now?

Look back at **What Do You Think?** on page 143. How have your answers and explanations changed?

Arranging Tables

Kate's class is in charge of arranging tables for a games day in the gym. The tables are shaped like red pattern blocks.

Students will play games in groups of 8, 10, and 12. Kate follows these rules for arranging the tables:

- Only 1 student can sit at each side of a table.
- Each student must have a place to sit.
- Tables must be joined along their sides, with sides of the same length together.
- The table arrangements should have at least 1 line of symmetry.

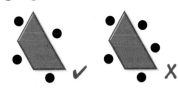

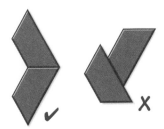

How can you arrange the tables to meet all the conditions?

A. Create table arrangements to seat groups of 8, 10, and 12 people.

B. Draw diagrams of the table arrangements. Label the lines of symmetry.

C. Can you find another arrangement for each group?

D. Describe how you found your solutions.

Multiplication and Division Facts

GOALS

You will be able to

- use mental math to solve multiplication and division problems
- relate division to multiplication
- identify patterns in a multiplication table
- solve problems by working backwards

There are 4 paddlers in each of 4 boats. How can you calculate the number of paddlers in the race? Show 2 ways.

Getting Started

Library Tables

Each day at noon, Daniella and her friends sit at tables in the library. The tables have equal numbers of students.

The chart shows what happened in one school week.

Number of Students at Tables

Day	Number of students in library	Number of tables being used	Number of students sitting at each table
Monday	12	3	▪
Tuesday	16	▪	4
Wednesday	25	5	▪
Thursday	▪	4	5
Friday	6	▪	6

How can you complete the chart?

A. How can you use counters to figure out the number of students at each table on Monday?

B. What calculation can you use to solve the problem in Part A? Explain.

C. How many tables did the students sit at on Tuesday? Use counters to determine the number.

D. What calculation can you use to solve the problem in Part C? Explain.

E. What are the missing numbers in the chart for Wednesday, Thursday, and Friday? Explain what you did to figure out each number.

What Do You Think?

Do you *agree* or *disagree* with each statement?
Explain your thinking.

1. When you multiply 2 whole numbers, the answer is always greater than both numbers.

2. There is more than one way to calculate 5 × 4.

3. The order doesn't matter when you divide one number by another number. For example,
$8 \div 2 = \blacksquare$ is the same as $2 \div 8 = \blacksquare$.

4. You can use this array of counters to represent both a multiplication situation and a division situation.

Multiplying by Skip Counting

You will need
- counters
- a number line

GOAL

Use skip counting to multiply.

Ken's community is collecting and donating used bicycles for children in Africa. The pictograph shows the number of bikes donated by 4 schools.

Bike Donations by School

Westside

Lakeside

Mountain View

Riverside

Each means 5 bikes.

How many bikes did each school donate?

Ken's Multiplication

There are 8 wheels for Mountain View School.
Each wheel represents 5 bikes.
I can use addition or multiplication to show
8 groups of 5 bikes.

factor

Any one of the numbers you multiply

product

The result when you multiply

$$2 \times 6 = 12$$

factors   product

$5 + 5 + 5 + 5 + 5 + 5 + 5 + 5 = $ ▨

$8 \times 5 = $ ▨

In the multiplication, 8 and 5 are the **factors**.

I'll skip count by 5s to calculate the **product** of 8×5.

A. How can you skip count by 5s to calculate 8×5?

B. How do you know you can write $9 \times 5 = $ ▨ to represent the number of bikes donated by Riverside School?

C. Complete the multiplication equation in Part B. Explain.

D. Write and complete multiplication equations to represent the number of bikes donated by Westside School and Lakeside School.

E. How many bikes were donated by each school? Explain.

Reflecting

F. How are Ken's addition equation and multiplication equation alike? How are they different?

G. What number patterns do you notice when you skip count by 5s?

Checking

1. The pictograph shows the number of bikes donated by 3 schools.

Bike Donations by School

Eastside 🛞🛞🛞🛞🛞🛞🛞

Gully View 🛞🛞🛞🛞🛞

Beachside 🛞🛞🛞🛞🛞🛞🛞🛞🛞

Each 🛞 means 2 bikes.

a) Write an addition equation and a multiplication equation to show the number of bikes donated by each school.

b) How many bikes were donated by each school? Explain how you calculated one answer.

Practising

2. Maya and Rey calculated the number of days in 5 weeks. They wrote $5 \times 7 = $ ▪.
Rey used a number line.

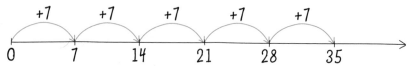

Maya counted: 5, 10, 15, 20, 25, 30, 35.

a) Which way of calculating 5×7 do you think is easier? Explain.

b) How many days are in 5 weeks?

c) How can you figure out the number of days in 6 weeks?

3. How many wheels are on 6 tricycles? Show your work.

4. Barrett skip counted on a number line to multiply.

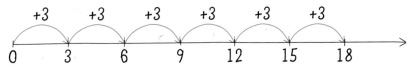

a) What 2 numbers did Barrett multiply? Write an equation.

b) How can Barrett extend his number line to calculate 7×3?

5. How many legs are on 3 spiders? Each spider has 8 legs. Show your work.

6. What is the missing number in each equation?

a) $2 \times 6 = \blacksquare$

b) $5 \times \blacksquare = 30$

c) $\blacksquare = 6 \times 8$

d) $\blacksquare \times 7 = 21$

e) $4 \times 5 = \blacksquare$

f) $6 \times \blacksquare = 36$

7. How many legs are on 6 bees?

8. a) Show how to calculate 9×1.

b) Show how to calculate 9×0.

c) Choose another 1-digit number and multiply it by 1. Why is it easy to multiply any number by 1?

d) Choose another 1-digit number and multiply it by 0. Why is it easy to multiply any number by 0?

9. Maddy skip counted by a 1-digit number until she reached 12. What numbers do you think she was multiplying? Write multiplication equations to show 2 answers.

10. When you skip count to multiply 2 numbers, how do you know which number to count by? How do you know when to stop counting? Explain, using an example.

Chapter 6

Lesson 2

Building on Multiplication Facts

You will need
* counters

GOAL

Skip count from facts you know to calculate other multiplication facts.

Four Hugs a Day

Nobody gets enough hugs a day
'Cause the minimum number is four
Now if you haven't got
Four Hugs today
Then you better get some more.

Annie's younger sister heard Charlotte Diamond from British Columbia sing "Four Hugs a Day" at a children's festival.

? How many hugs should Annie's sister get in 1 week?

Annie's Multiplication

$7 \times 4 = \blacksquare$ represents the number of hugs in 1 week (7 days).

I'll use counters to show 7 fours.

$7 \times 4 = \blacksquare$

$5 \times 4 = 20$

$7 \times 4 = 4 + 4 + 4 + 4 + 4 + 4 + 4$

$5 \times 4 = 20$

I can skip count by 4s from 20 to get 7 fours.

174

NEL

A. Complete Annie's multiplication. Explain what you did.

B. How many hugs should Annie's sister get in 1 week?

Reflecting

C. How can you use your answer to $7 \times 4 = \blacksquare$ and skip counting or adding to calculate 8×4 and 9×4? Explain, using counters.

Checking

$7 \times 6 = \blacksquare$

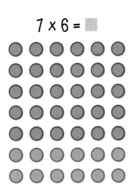

1. Shani wants 6 hugs a day. She used counters to calculate the number of hugs in 1 week.
   a) Why do you think Shani arranged the counters into these groups?
   b) Complete Shani's multiplication. Show your work.
   c) Show another way to calculate 7×6 with counters. Sketch what you did.

Practising

2. a) How many wheels are on 6 inline skates?
   b) How many wheels are on 8 inline skates? Use your answer to part (a).

3. Use one fact to help you complete the other fact. Explain what you did.
   a) $6 \times 6 = 36$, so $7 \times 6 = \blacksquare$
   b) $8 \times 5 = 40$, so $9 \times 5 = \blacksquare$
   c) $7 \times 4 = 28$, so $7 \times 6 = \blacksquare$
   d) $5 \times 6 = 30$, so $5 \times 7 = \blacksquare$

4. Explain how to use other multiplication facts to calculate 9×8.

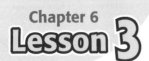
Doubling Multiplication Facts

You will need
• counters

GOAL

Double facts you know to calculate other multiplication facts.

Some students are learning Irish dances. The Reel of Four is a dance for groups of 4. The Reel of Eight is a dance for groups of 8.

? **What is the total number of dancers if there are 3 groups in each dance?**

double

Add a number to itself

For example, double 28 is 28 + 28 = 56.

Alec's Multiplication

The counters show 3 groups of 4 dancers for the Reel of Four.

3 × 4 = 12

3 x 4

I can **double** the number of counters in each group to show 3 groups of 8 dancers, or 3 × 8.

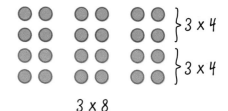

3 x 4

3 x 4

3 x 8

I can double the fact 3 × 4 = 12 to figure out 3 × 8.

A. How can you double 3 × 4 = 12 to figure out 3 × 8?

B. How many dancers are in 3 groups of 8 dancers?

Reflecting

C. Multiply 7 by 4. How can you use the answer to multiply 7 by 8?

Checking

1. a) How many dancers are in 6 groups of 2? How can you use your answer to calculate the number of dancers in 6 groups of 4?

b) How many dancers are in 7 groups of 2? How can you use your answer to calculate the number of dancers in 7 groups of 4?

Practising

2. a) How many dots are shown on the dice?

b) How many dots are on 8 dice if each die shows 6 instead of 3? Explain how you can use your answer to part (a) to calculate.

3. How can you use doubling to complete the 4× and 8× rows of this table?

×	5	6	7	8	9
2	10	12	14	16	18
4					
8					

> I can multiply any number by 8 by doubling the number 3 times.

4. How can you use doubling to complete the 6× row?

×	5	6	7	8	9
3	15	18	21	24	27
6					

5. Alec has a strategy for multiplying by 8.

a) Multiply 5 by 8 using Alec's method. Try other numbers.

b) Explain why his method works.

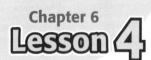
Halving and Doubling Multiplication Facts

You will need
- counters

GOAL

Halve and double facts you know to calculate other multiplication facts.

Adrian's school is getting a climber made from used tires. It will have 9 rows of 5 tires.

178

How many tires are needed to build the climber?

Diane's Strategy

My array shows that 9×5 equals $8 \times 5 + 1 \times 5$.

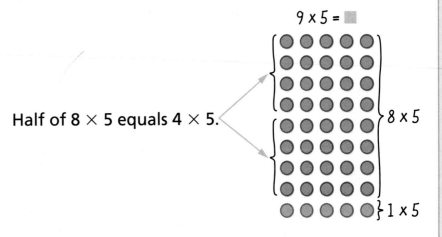

$9 \times 5 = \blacksquare$

8×5

1×5

Half of 8×5 equals 4×5.

I can double $4 \times 5 = 20$ to figure out 8×5.

$9 \times 5 = 8 \times 5 + 1 \times 5$
$9 \times 5 = 4 \times 5 + 4 \times 5 + 1 \times 5$
$9 \times 5 = 20 + 20 + 5$
$9 \times 5 = 45$

We need 45 tires to build the climber.

Reflecting

A. How do you think Diane knew that 4×5 is half of 8×5?

B. How can you use Diane's halving and doubling strategy to calculate 7×5?

Checking

1. A tire climber has 9 rows of 6 tires.
   a) What multiplication equation represents the number of tires?
   b) Model the climber using rows of counters.
   c) Split 8 rows in half. Explain how to use the 2 halves and the extra row to calculate the total number of tires.
   d) Calculate the number of tires another way.

Practising

2. a) How many batteries are in 6 packages of 4 batteries?
   b) How many batteries are in 6 packages of 8 batteries?
   c) How many batteries are in 7 packages of 8 batteries? Show how you can use your answer to part (b) to figure it out.

3. Calculate.
   a) $4 \times 9 = $ ▪
   b) ▪ $= 3 \times 6$
   c) $5 \times 6 = $ ▪
   d) ▪ $= 9 \times 6$
   e) ▪ $= 8 \times 8$
   f) $9 \times 8 = $ ▪

4. Complete each equation.
   a) $4 \times $ ▪ $= 24$
   b) $6 \times $ ▪ $= 42$
   c) $28 = $ ▪ $\times 7$
   d) $32 = $ ▪ $\times 8$

5. Calculate the first product. How can you use the answer to calculate the second product?
   a) $2 \times 8 = $ ▪, so $5 \times 8 = $ ▪
   b) $3 \times 7 = $ ▪, so $7 \times 7 = $ ▪
   c) $3 \times 4 = $ ▪, so $7 \times 4 = $ ▪
   d) $3 \times 5 = $ ▪, so $5 \times 5 = $ ▪

6. Choose a 1-digit number and multiply it by 3. How can you use the answer to multiply your number by 6? by 7?

Lesson 5

Using 10s to Multiply

GOAL

Multiply by 10 to calculate other facts.

Some students in Jade's class used fingers to model multiplication. First they modelled 6 tens or 6 × 10.

They modelled 6 × 5.

They modelled 7 × 5.

? **How can you use counting by 10s to calculate other multiplication facts?**

Lesson 6

Multiplying by 8 and 9

You will need
- 10-frames
- counters

GOAL

Multiply by 10 to multiply by 8 and 9.

Michael's school rented 8 ten-passenger vans to take students on field trips. On the first trip, each van carried 8 students. On the second trip, each van carried 9 students.

 How many students went on each field trip?

Michael's Multiplication

8 counters show the number of students in each ten-passenger van.
I need to calculate 8 × 8 to figure out the total number of students on the first field trip.

8 × 10 = 80
I can take away 8 twos, or 16, from 80 to calculate 8 × 8.

Hailey's Patterns

Each van carries 9 students on the second trip.
There are 8 vans.
8 × 9 is the same as 9 × 8. I'll calculate 8 × 9 using
patterns in a multiplication table.

×	1	2	3	4	5	6	7	8	9
9	9	18	27	36					

I see a pattern in the tens digits of each 2-digit
product.
I also see a pattern when I add the digits of each
2-digit product.

A. Complete Michael's multiplication.

B. How can Michael use his 10-frames to calculate
8 × 9?

C. Complete Hailey's multiplication table. What
patterns do you see in the table?

D. Use the patterns to calculate 8 × 9. Show your work.

E. How many students went on each field trip?

Reflecting

F. Why do you think Michael subtracted 8 twos
from 80?

G. How can you use multiplying by 10 to
calculate 5 × 9? Explain, using Michael's
10-frames or Hailey's patterns.

Checking

1. Another school rented 7 ten-passenger vans. On the first field trip, each van carried 8 students. On the second field trip, each van carried 9 students. How many students went on each trip? Solve each problem using 10-frames.

Practising

2. Mandy's school has 6 hopscotch games like this one. What is the total number of squares painted on the playground?

3. How can you use the 10× row shown below to complete the 8× row and the 9× row?

×	4	5	6	7	8	9
8	32					
9	36					
10	40	50	60	70	80	90

4. How can you use these tens to calculate 4×9 and 4×8?

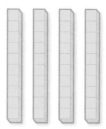

5. At the Edmonton Heritage Festival, a quesadilla costs 4 food tickets.
   a) What is the total cost of 10 quesadillas?
   b) How can you use your answer in part (a) to figure out the cost of 9 quesadillas?

6. Annie said, "It's easy to multiply a 1-digit number by 9." Why do you think she said that the 9× facts are easy to remember?

184

Finger Multiplication

Diane's Finger Calculating

I numbered my fingers from 1 to 10. I can use my fingers to multiply any 1-digit number by 9.

If you multiply a number by 9, you put down the finger for that number. You count the number of tens on one side of that finger, and the number of ones on the other side.

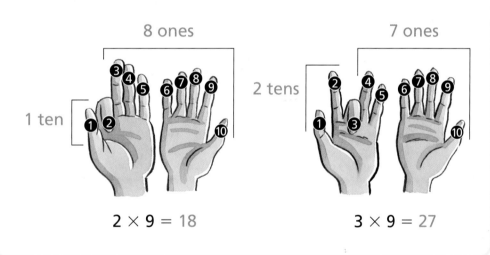

$2 \times 9 = 18$ $3 \times 9 = 27$

1. Use your fingers to multiply 9 by 4, 5, 6, 7, 8, and 9. Draw sketches to show what you did.

2. Explain why the finger method works. Hint: Compare the finger method to a number pattern you found in Lesson 6.

Mid-Chapter Review

Frequently Asked Questions

Q: What are some ways to multiply numbers?

A1: You can skip count from 0 by one of the factors.
For example, you can skip count by 5s from 0 to
calculate 8×5. You stop when you count 8 fives.

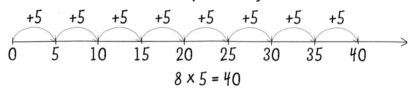

$8 \times 5 = 40$

A2: You can also add to or subtract from a fact that
you know. For example, if you know $8 \times 5 = 40$,
you can add 5 to 40 to calculate $9 \times 5 = 45$. You
can subtract 5 from 40 to calculate $7 \times 5 = 35$.

A3: You can use halving and doubling to calculate
with facts that you know. For example, 7×5 or
7 fives can be split up as shown.
$7 \times 5 = 3$ fives $+ 3$ fives $+ 1$ five
$7 \times 5 = 15 + 15 + 5$
$7 \times 5 = 35$

$3 \times 5 = 15$ $3 \times 5 = 15$ $1 \times 5 = 5$

A4: You can multiply by 10 and then subtract to
multiply by other numbers close to 10. For
example, you can use the fact that 4 tens equal 40
to calculate 4×9.
$40 - 4 = 36$, so $4 \times 9 = 36$.

Practice

Plant Height Each Week

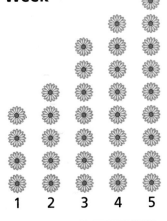

1   2   3   4   5

Each ❀ means 2 cm.

Lesson 1

1. Olya planted a sunflower because it is the national flower of Ukraine, the country of her grandparents. This pictograph shows the height of the sunflower as it grows. Calculate the heights for each week using multiplication.

2. How can you use skip counting to calculate 5×8?

Lesson 2

3. Nicky skip counted by 5s from 0 to 35.
   a) What number did he multiply by 5? Explain.
   b) How can he skip count from 35 to multiply 9 by 5?

Lesson 3

4. Marney's mother bought some Vancouver 2010 Canadian flag pins for gifts. Each pin cost $8.
   a) What is the cost of 3 pins? Show your work.
   b) What is the cost of 6 pins? How can you use your answer from part (a) to find out?

Lesson 4

5. Calculate the first product. How can you use the answer to calculate the second product?
   a) $2 \times 8 = $ ▨, so $4 \times 8 = $ ▨
   b) $2 \times 9 = $ ▨, so $5 \times 9 = $ ▨
   c) $3 \times 8 = $ ▨, so $6 \times 8 = $ ▨
   d) ▨ $= 9 \times 3$, so ▨ $= 9 \times 7$

Lesson 6

6. Calculate. Explain what you did for one answer.
   a) 5×8 b) 8×7 c) 2×7 d) 3×9

7. What is the missing number in each equation?
   a) $7 \times 0 = $ ▨
   b) $5 \times $ ▨ $= 5$
   c) ▨ $\times 9 = 0$
   d) ▨ $= 0 \times 8$

Lesson 7 — Sharing and Grouping

You will need
- counters
- a number line

GOAL

Use 2 meanings of division to solve problems.

Diane created 2 division problems about pouring vegetable scraps into worm-composting tubs. The worms in the tubs change the scraps into soil.

Problem 1
We poured 40 small pails of vegetable scraps equally into 8 tubs. How many pails of vegetable scraps went into each tub?

Problem 2
We poured 40 small pails of vegetable scraps equally into some tubs. Each tub contained 5 pails of vegetable scraps. How many tubs were there?

❓ How can you solve Diane's division problems?

Ken's Solution

Problem 1 is a **sharing** problem because 40 pails are shared equally among 8 tubs.

$$40 \div 8 = \blacksquare$$

total pails   number of tubs   number of pails in each tub

I'll use counters to model 40 pails divided into 8 equal groups. That will tell me the number of pails in each tub.

$$40 \div 8 = 5 \qquad 8\overline{)40}^{5}$$

Each tub took 5 pails of scraps.

Annie's Solution

Problem 2 is a **grouping** problem because 40 pails are placed in groups of 5 pails.

$$40 \div 5 = \blacksquare$$

total pails   number of pails   number of
in each tub   tubs

I'll keep subtracting 5 pails until I get to 0 to figure out the number of tubs.

-5   -5   -5   -5   -5   -5   -5   -5

0   5   10   15   20   25   30   35   40

There are 8 groups of 5 in 40.

$$40 \div 5 = 8 \qquad 5\overline{)40}^{8}$$

Diane's family has 8 tubs.

Reflecting

A. How do you think each student would use counters to calculate $18 \div 2$?

Checking

1.

> **Problem 1**
> Hari's family composted 42 kg of scraps in 7 weeks. How many kilograms of scraps did they compost each week?

> **Problem 2**
> Hari's family composted 6 kg of scraps each week. How many weeks did it take them to compost 42 kg?

  a) Problem 1 is a sharing problem. Solve it.

  b) Problem 2 is a grouping problem. Solve it.

Practising

2. How many groups of 6 students are in a class of 24 students? Show your work.

3. How many students are on 7 equal teams in a class of 35 students? Show your work.

4. Calculate.

  a) $6\overline{)18}$ **b)** $3\overline{)21}$ **c)** $8\overline{)48}$ **d)** $4\overline{)28}$

5. What is the missing number in each equation?

  a) $24 \div 6 = \blacksquare$ **b)** $45 \div \blacksquare = 5$

6. a) How does this number line help you divide 28 by 4?

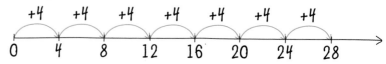

  b) How can you use a number line to calculate $35 \div 5$?

7. Create a problem that can be solved by dividing 56 by 8. Explain how to solve the problem.

8. How can you use adding or subtracting by a number to divide? Use the example $35 \div 5$ to help you explain.

Comparing Products

You will need
- number cards
  0 to 10 (2 sets)

Number of players: 2 to 4
How to play: Multiply numbers on cards.
Score points for products.

- Step 1 Deal 2 cards to each player.

- Step 2 All players multiply the numbers on their cards.

- Step 3 Each player calculates a score:
  1 point if the product is an odd number
  1 point if the product is greater than 50

Repeat until a player reaches a total of 10 points.

Diane's Turn

I multiplied 9 by 7 and got 63.
I scored 2 points because 63
is an odd number and is greater
than 50.

9 7

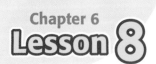

Chapter 6
Lesson 8

Division and Multiplication

You will need
* counters

GOAL

Divide by using related multiplication facts.

35 students plan to play the Aboriginal game Turtle Catcher. They want to play in equal groups of 5 students. Each group will play in a different circle.

Turtle Catcher
* 1 turtle catcher and 4 turtles get inside a circle.
* After saying "Go," the turtle catcher tries to tag all turtles. Tagged turtles must leave the circle.
* Turtles are safe if they lie on their backs with their arms and legs raised (for up to 6 seconds).

How many circles are needed for the students to play the game?

192

dividend

The starting number in a division operation

divisor

The number you divide by in a division operation

quotient

The whole number result you get when you divide

$72 \div 9 = 8$

dividend | quotient

divisor

divisor dividend

Michael's Multiplication

I'll use an array and write 2 equations to figure out the number of groups of 5 in 35.

$35 \div 5 = \blacksquare$
$\blacksquare \times 5 = 35$
Both equations describe the array.

In the division equation, 35 is the **dividend** and 5 is the **divisor**.
The answer for ▨ is the **quotient**.
The dividend is also the product in the multiplication equation.
The 2 factors in the multiplication equation are the divisor and quotient in the division equation.

A. How can you use Michael's array to figure out the dividend, divisor, and quotient when 35 is divided by 5?

B. How many circles are needed for the students to play the game?

Reflecting

C. How can you use multiplication to check the quotient when you divide? Use $28 \div 7$ to explain.

Checking

1. 45 students want to play Turtle Catcher in equal groups of 5. How many circles do they need? Show your work.

Practising

2. Barrett used 32 counters to make this array. What pairs of numbers will he probably write in each equation?

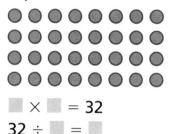

$$\blacksquare \times \blacksquare = 32$$
$$32 \div \blacksquare = \blacksquare$$

3. Calculate.
   a) $18 \div 2 = \blacksquare$
   b) $\blacksquare = 81 \div 9$
   c) $64 \div 8 = \blacksquare$
   d) $7\overline{)56}$
   e) $8\overline{)72}$
   f) $7\overline{)49}$

4. 36 students are going on a guided nature walk at the Kerry Wood Nature Centre in Red Deer, Alberta. They must have at least 1 adult supervisor for every 6 students. How many adults are needed? Show your work.

5. A beater for a bodhran drum costs $5. How many beaters can be bought for $40? Show your work.

6. A class of 36 students is buying packs of juice. Each pack contains 6 juice boxes. How many packs do they need to buy so that every student can have 1 juice box with no juice boxes left over?

7. Lukina is using a number line to divide.

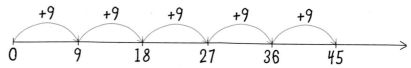

a) Write an equation for Lukina's division.
b) How could she have used multiplication instead of a number line to divide the numbers?

8. a) Choose any 3 one-digit numbers greater than 1. How can you use multiplication to divide each number by 1?
b) Why is it easy to divide any number by 1?

9. a) Complete each division using multiplication. Show your work.

$0 \div 2 = \blacksquare$ $0 \div 4 = \blacksquare$ $0 \div 8 = \blacksquare$
$\blacksquare \times 2 = 0$ $\blacksquare \times 4 = 0$ $\blacksquare \times 8 = 0$

b) Why is it easy to divide 0 by any number greater than 0?

10. Check each division for errors using multiplication. Show your work.

a) $3\overline{)27}$ with 9 above   b) $6\overline{)54}$ with 8 above   c) $8\overline{)64}$ with 8 above   d) $9\overline{)81}$ with 8 above

11. What number is missing in each calculation?

a) $45 \div \blacksquare = 9$

b) $\blacksquare \div 5 = 5$

c) $\blacksquare = 72 \div 8$

d) $7\overline{)\blacksquare}$ with 9 above

e) $4\overline{)36}$ with $\blacksquare$ above

f) $\blacksquare\overline{)49}$ with 7 above

12. How can you use the multiplication fact $6 \times 5 = 30$ to calculate $30 \div 5$ and $30 \div 6$?

Patterns in a Multiplication Table

You will need
- a multiplication table

GOAL

Use number patterns in a table to multiply and divide.

Alec identified number patterns in a multiplication table. He used the patterns to calculate multiplication and division facts.

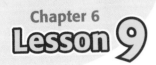

Alec's Patterns

I can divide 54 by 9 by looking for 54 in the 9× row.

×	0	1	2	3	4	5	6	7	8	9
9	0	9	18	27	36	45	54	63	72	81

The 0× and 1× facts are easy to remember.

×	0	1	2	3	4	5	6	7	8	9
0	0	0	0	0	0	0	0			
1	0	1	2	3	4	5	6			

How can you use patterns in a multiplication table to multiply and divide?

Matching Pairs

Number of players: 2 or more

How to play: Match 2 cards that show the same answer.

You will need

- Matching Pairs Game Cards (blackline master)

- **Step 1** Shuffle the cards and place them face down in an array of 6 rows of 4.

- **Step 2** Decide who goes first. Player 1 turns over 2 cards. If the cards have the same missing number, Player 1 keeps the cards and turns over another 2 cards. If the cards don't have the same answer, the player turns the cards back in the same position.

- **Step 3** Take turns until all the cards have been taken.

The player with the most cards is the winner.

Hailey's Turn

I turned over 2 cards.
The missing number on both cards is 6.
I keep these 2 cards and turn over 2 more cards.

$$36 \div \blacksquare = 6$$

$$\blacksquare \times 8 = 48$$

Solving Problems by Working Backwards

GOAL

Work backwards to solve problems.

Desmond sorted some baseball cards into 8 equal groups. Then he sorted each group into 2 piles of 3 cards each.

? **How many baseball cards did Desmond start with?**

Cory's Solution

Understand the Problem
I need to figure out how many baseball cards Desmond had to start with.

Make a Plan
I'll use a diagram to represent each step of the problem.

| Number of baseball cards? | → | Divide by 8. | → | Divide by 2. | → | 3 cards in each small pile. |

I'll work backwards to solve the problem.

Carry Out the Plan

My new diagram shows the same steps of the problem in reverse order.

| Number of baseball cards? | ← | Multiply by 8. | ← | Multiply by 2. | ← | 3 cards in each small pile. |

Desmond had $3 \times 2 \times 8 = 48$ cards to start with.

Reflecting

A. How did Cory use his first diagram to make the second diagram?

B. How can you check Cory's answer?

Checking

1. Taylor sorted some stickers into 7 equal groups. Then she sorted each group into 4 piles of 2 stickers.
   a) How many stickers did she have altogether? Draw a diagram to show the steps of the problem.
   b) Work backwards to solve the problem.

Practising

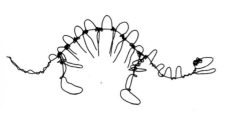

2. Ryan is making a wire sculpture. He cut a wire into 4 equal pieces. Then he cut 2 cm from each piece so each piece was 7 cm long. How long was the wire he started with? Show your work.

3. Juliette's grandfather gave her $20. She spent $5 and put the rest in her piggy bank. Now she has $30 in her bank. How much money was in her bank before her grandfather gave her money?

4. Create your own working backwards problem.

Chapter Review

Frequently Asked Questions

Q: **What are some ways to divide?**

A1: You can use repeated addition or repeated subtraction. For example, to calculate 45 ÷ 5, show jumps on a number line:

Repeated Addition

+5 +5 +5 +5 +5 +5 +5 +5 +5

0   5   10   15   20   25   30   35   40   45

9 jumps of 5 to 45

9
5)45 and 45 ÷ 5 = 9

Repeated Subtraction

−5 −5 −5 −5 −5 −5 −5 −5 −5

0   5   10   15   20   25   30   35   40   45

9 jumps of 5 to 0

9
5)45 and 45 ÷ 5 = 9

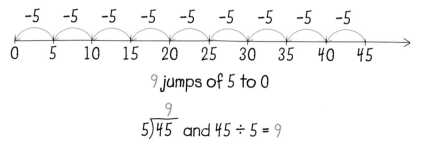

4 × 5 = 20
20 ÷ 4 = 5

A2: You can use multiplication. For example, to calculate 20 ÷ 4 = ■, think of a number you can multiply 4 by to get 20. That number is the quotient.
5 × 4 = 20, so 20 ÷ 4 = 5.

Practice

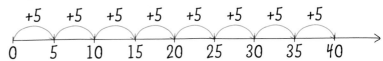

Lesson 1

1. **a)** The number line shows skip counting to multiply. What 2 numbers are being multiplied? Explain.

    +5   +5   +5   +5   +5   +5   +5   +5

    0   5   10   15   20   25   30   35   40

   b) Show another way to multiply these 2 numbers on a number line.

Lesson 2

2. Use one fact to complete the other fact. Explain what you did.

    a) 5 × 5 = 25, so 5 × 6 =
    b) 7 × 5 = 35, so 7 × 6 =
    c) 5 × 9 = 45, so 4 × 9 =

3. Malik's class bought 6 packages of pens.

    a) How many pens did they buy?
    b) Explain how you can use your answer to calculate the number of pens in 7 packages.

Lesson 3

4. Jordan's grandmother is making a bracelet.

    a) What multiplication can you use to represent the number of beads?
    b) How can you use the array to calculate the number of beads in 3 rows of 8?

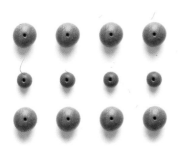

Lesson 4

5. It costs 5 cents each minute for Tamara to phone Calgary.

    a) How much will a 4-minute phone call cost?
    b) Use your answer to calculate the cost of a 9-minute phone call. Explain what you did.

6. Calculate the first product. How can you use this product to calculate the second product?
   a) $3 \times 7 =$, so $6 \times 7 =$
   b) $4 \times 6 =$, so $9 \times 6 =$

Lesson 6

7. a) A lacrosse team has 10 players. How many players are on 7 lacrosse teams?
   b) A softball team has 9 players. How many players are on 7 softball teams? Use your answer to part (a).

Lesson 7

8. Aleeza's class of 28 students is learning about India. Each group of 4 students will make a sari. How many groups can be formed?

9. Chi has a $40 gift certificate. How many used books can she buy if each book costs $5?

Lesson 8

10. Use multiplication to calculate each quotient.
    a) $72 \div 9$ b) $48 \div 6$ c) $28 \div 7$ d) $54 \div 9$

11. What is the missing number in each calculation?
    a) ▨ $\times 1 = 7$ c) ▨ $= 6 \times 1$
    b) $3 \times$ ▨ $= 0$ d) $8 \times$ ▨ $= 8$

Lesson 10

12. 28 people camped in 4 tents. The smallest tent had 4 people. The other tents each had the same number of people. How many people were in each tent?

What Do You Think Now?

Look back at **What Do You Think?** on page 169. How have your answers and explanations changed?

Chapter Task

Task Checklist

✔ Did you show all your steps?
✔ Did you explain your thinking?
✔ Did you show enough detail?

Planning a Bone Puzzle Game

In the Inuit game called *Inukat* or Bone Puzzle, players are given equal numbers of bones. The winner is the first one to use the bones to make the shape of a seal flipper.

Suppose a game uses between 60 and 85 bones and follows all of these rules:
- All players get the same number of bones with no bones left over.
- No player can have more than 9 bones.
- There must be from 2 to 9 players.

❓ How many players can play Inukat?

A. How do you know that the game must be played with 7 or more players?

B. How do you know that each player must have 7 or more bones?

C. Calculate the possible number of bones and possible number of players. Explain your strategy.

NEL

Fractions and Decimals

GOALS

You will be able to

- name and represent fractions of a region and fractions of a set

- describe how fractions can be used

- compare and order fractions

- describe and represent decimals

- add and subtract decimals, including money amounts

- solve problems by drawing diagrams

What fractions do you see in the kite?

Getting Started

Tile Fractions

Aneela is making designs with square tiles.

? What fractions can you show with square tiles?

NEL

A. How does  show $\frac{3}{4}$?

B. How does  show $\frac{1}{4}$?

C. What does the 4 in the fraction $\frac{3}{4}$ tell about the red and yellow tile design?

D. What does the 3 in the fraction $\frac{3}{4}$ tell about the red and yellow tile design?

E. What fractions does 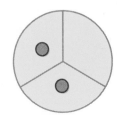 show?

F. Make 3 more tile designs that show fractions. Write the fractions.

What Do You Think?

Do you *agree* or *disagree* with each statement? Explain your thinking.

1. This picture shows $\frac{3}{4}$.

2. This picture shows $\frac{2}{3}$.

3. Design B is half of Design A.

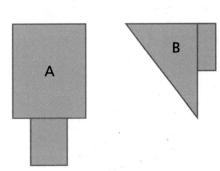

Fractions of a Whole

You will need
- pencil crayons
- Fraction Shapes (blackline master)

GOAL

Name and represent fractions of a whole.

Joshua's rock-climbing team is called the *Numerators.* Each person's T-shirt shows a different fraction. Joshua's T-shirt shows $\frac{1}{6}$.

? **What pictures might Lang, Tien, and Hailey put on their T-shirts?**

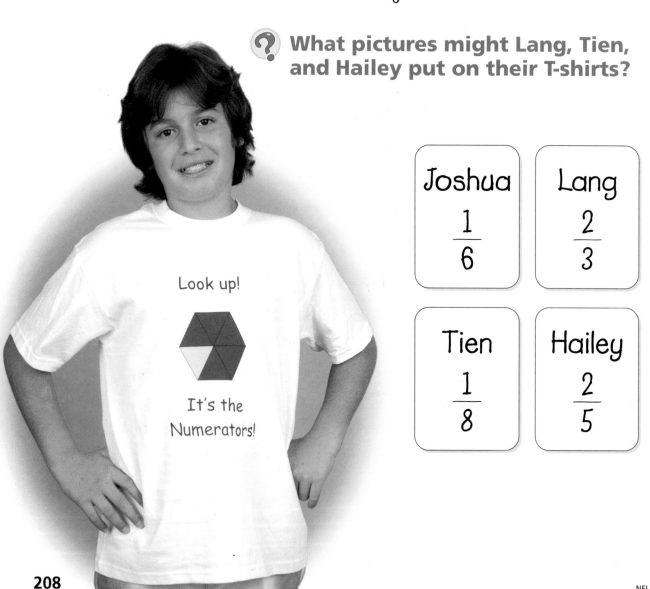

Look up!

It's the Numerators!

| Joshua $\frac{1}{6}$ | Lang $\frac{2}{3}$ |
| Tien $\frac{1}{8}$ | Hailey $\frac{2}{5}$ |

NEL

Lang's Fraction

I'll use a rectangle for the whole.

My fraction is $\frac{2}{3}$. That's *two thirds.*

It has 3 in the **denominator**, so I need 3 equal parts.

It has 2 in the **numerator**, so I'll use blue for 2 of the parts.

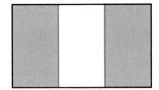

denominator

The number below the bar in a fraction. It tells the number of equal parts in one whole.

For example, this whole is divided into 4 equal parts.

$\frac{1}{4}$

numerator

The number above the bar in a fraction. It tells the number of equal parts the fraction represents.

For example, this fraction tells about 1 of the equal parts.

$\frac{1}{4}$ ←

A. Why do you think Lang needed 3 *equal* parts?

B. Colour another picture that shows $\frac{2}{3}$. How do you know it shows $\frac{2}{3}$?

C. Colour 2 different T-shirt pictures for Tien and 2 for Hailey.

Reflecting

D. How did you decide what whole shape to use each time?

E. How does each picture really show 2 fractions?

F. How do you know it is always possible to use different pictures to show the same fraction?

Checking

1. **a)** Colour a picture to show three eighths.
   b) What other fraction does your picture show?

Practising

2. Hannah coloured these fraction pictures.

a) Which picture shows $\frac{1}{4}$? How do you know?

b) What fractions do the other pictures show?

3. **a)** Colour a picture to show any fraction. Write the fraction you coloured.
   b) What other fraction does your picture show?

4. Use a fraction to describe how full each glass is.

a)

b)

5. Use a rectangle divided into twelfths.
   a) Colour your rectangle so more than $\frac{4}{12}$ is yellow, less than $\frac{4}{12}$ is blue, and the rest is green.
   b) What fraction of your rectangle is green?

6. **a)** Colour a picture to show $\frac{7}{10}$.
   b) What other fraction does your picture show?

7. Which pictures *do not* show $\frac{5}{6}$? Explain how you know.

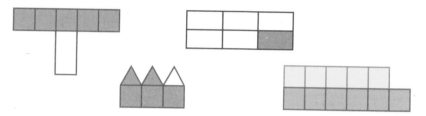

8. Annie ate $\frac{1}{2}$ of her apple and Desmond ate $\frac{1}{2}$ of his. Desmond ate more apple than Annie. How is that possible?

9. How does this picture show each fraction?

a) $\frac{4}{10}$ c) $\frac{10}{10}$

b) $\frac{7}{10}$ d) $\frac{0}{10}$

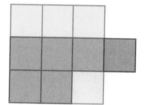

10. What other fractions do you see in the picture in Question 9? Give at least 3 examples. Explain what each fraction shows about the picture.

11. What fractions describe each picture?

a)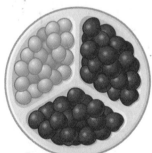

b)

12. Look for something in your classroom that's divided into equal parts. Sketch what you found. Describe the fractions you see in your object.

Fractions of a Group

> **GOAL**
>
> Name, represent, compare, and order fractions of a group.

Cole's family took pictures of some animals they saw on their fishing trip.

polar bear

caribou

ermine

ptarmigan

musk ox

? **How can you use fractions to describe the animals in this group?**

Cole's Description

4 of the 5 animals have fur.
I can say *four fifths*.
I can write $\frac{4}{5}$.

Does it have fur?
✓ polar bear
✓ musk ox
✓ ermine
✓ caribou
X ptarmigan

A. What fraction tells about the animals with feathers?

B. Which fraction is greater, $\frac{4}{5}$ or the fraction that tells about feathers? How do you know?

C. Use 3 more fractions to describe the animal pictures.

Reflecting

D. Put your fractions from Part C in order from least to greatest. How can you show that the order is right?

E. The bears to the left show $\frac{2}{3}$, but the rectangle does not. How are parts of a group different from parts of a shape?

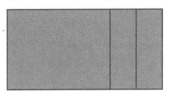

Checking

1. a) Write 3 different fractions that describe this group of animals.

seal

lemming

walrus

ermine

b) Put the 3 fractions in order from least to greatest.

Practising

2. The children in this picture are playing Turtle Catcher. The turtles have green vests.
   a) What fraction describes children who are safe?
   b) What other fractions tell about the game?

3. a) Model $\frac{2}{3}$ with things in your desk. Sketch a picture of your model. Circle the $\frac{2}{3}$.

   b) Make a new group that shows $\frac{3}{3}$. Sketch a picture.

4. a) Sketch a picture of a group of crayons. Make $\frac{2}{6}$ blue, $\frac{3}{6}$ green, and $\frac{1}{6}$ red.

   b) Put the fractions in order from least to greatest.

5. If 2 fractions have the same denominator, how can you tell which fraction is greater?

6. Luis eats $\frac{1}{2}$ of his strawberries, and Diane eats $\frac{1}{2}$ of hers. Do they eat the same number of strawberries? How do you know?

7. How are these 2 models for $\frac{1}{6}$ alike? How are they different?

214

Lesson 3

Sorting Fractions

GOAL

Describe how fractions are alike and different.

One of these fractions does not belong with the others.

$$\frac{3}{4} \qquad \frac{4}{4} \qquad \frac{3}{3}$$

Ken's Solution

I think $\frac{3}{4}$ does not belong.

Julia's Solution

I think $\frac{4}{4}$ does not belong.

Aneela's Solution

I think $\frac{3}{3}$ does not belong.

How can you use fraction shapes or counters to show that Ken, Julia, and Aneela are all correct?

Comparing and Ordering Fractions

You will need
- Fraction Strips (blackline master)

GOAL

Compare and order fractions with the same numerator and different denominators.

Tien, Olivia, and Annie are racing on a track. Tien is $\frac{1}{4}$ of the way, Olivia is $\frac{1}{3}$ of the way, and Annie is $\frac{1}{6}$ of the way.

? **Who is winning the race: Tien, Olivia, or Annie?**

Joshua's Model

I'll use fraction strips to model the whole track and the parts that Tien and Olivia have run.

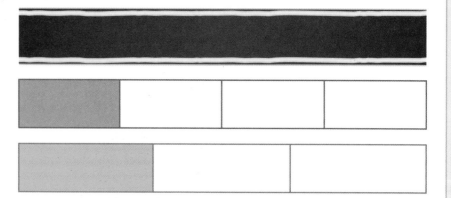

A. Which colour shows $\frac{1}{4}$? How do you know?

B. Which is more, $\frac{1}{3}$ or $\frac{1}{4}$? How do you know?

C. Use a fraction strip to show how far Annie has run.

D. Who is winning the race? How do you know?

Reflecting

E. When you compare 2 fractions that have 1 in the numerator, how can you tell which fraction is greater?

F. How can you use your answer to Part E to help you order other fractions of the track with the same numerator, such as $\frac{2}{3}$, $\frac{2}{4}$, and $\frac{2}{6}$?

Checking

1. Who is farther along the track?
   a) Ken is $\frac{1}{8}$ of the way, and Cory is $\frac{1}{5}$ of the way.
   b) Aneela is $\frac{1}{3}$ of the way, and Julia is $\frac{1}{5}$ of the way.
   c) Lang is $\frac{4}{5}$ of the way, and Alec is $\frac{4}{8}$ of the way.

2. Put these fractions of the track in order from least to greatest.

   $$\frac{1}{2} \qquad \frac{1}{5} \qquad \frac{1}{8} \qquad \frac{1}{3}$$

Practising

3. a) What fraction of the yellow block does the red block cover?
   b) What fraction of the yellow block does the blue block cover?
   c) Which fraction from parts (a) and (b) is greater? How do you know?

4. Draw a picture to show that $\frac{3}{4}$ of a loaf of bread is more than $\frac{3}{8}$ of it.

5. These fractions represent parts of a strip of ribbon. Put them in order from least to greatest.

$$\frac{4}{5} \qquad \frac{4}{10} \qquad \frac{4}{4} \qquad \frac{4}{8}$$

6. Is the following statement true or false? Explain your answer.

   $\frac{3}{5}$ is between $\frac{3}{4}$ and $\frac{3}{6}$.

7. a) Draw 2 identical squares. Colour $\frac{2}{5}$ of one square.

   b) Colour $\frac{2}{6}$ of the other square.

   c) Which fraction is greater, $\frac{2}{5}$ or $\frac{2}{6}$?

   d) Would $\frac{2}{8}$ of the square be greater or less than $\frac{2}{5}$? Explain how you know.

8. A diver checks his air tank before and after each dive. While the diver is underwater, some of the air in the tank gets used up. Each pair of fractions tells how full a diver's air tank was before and after a dive. Which is the "after" fraction in each pair?

   a) $\frac{2}{3}$ full or $\frac{2}{8}$ full

   b) $\frac{5}{6}$ full or $\frac{5}{8}$ full

9. Fill in the blank 3 different ways to make the sentence true.

   $\frac{5}{\blacksquare}$ of a glass of juice is less than $\frac{5}{6}$ of it.

10. Which of these fractions of the same whole thing are easiest to compare? Why?

    A. $\frac{2}{6}$ and $\frac{2}{8}$ B. $\frac{2}{7}$ and $\frac{3}{7}$ C. $\frac{2}{6}$ and $\frac{3}{7}$

11. Why doesn't it make sense to compare fractions when they are parts of different wholes? Use pictures and words to explain.

Food Fractions

Michelle's favourite food is a kind of meat pie called *tourtière*. She says that thinking about *tourtière* helps her understand fractions!

How can these pictures help Michelle compare $\frac{9}{10}$ and $\frac{11}{12}$?

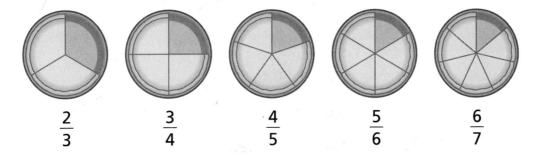

| $\frac{2}{3}$ | $\frac{3}{4}$ | $\frac{4}{5}$ | $\frac{5}{6}$ | $\frac{6}{7}$ |

1. What fraction tells about the largest amount of a pie that's left? What fraction tells about the smallest amount?

2. What fraction tells about the largest amount of a pie that's been eaten? What fraction tells about the smallest amount?

3. Copy and complete the chart to describe the pies.

Fraction that's left	$\frac{2}{3}$	$\frac{3}{4}$	$\frac{4}{5}$	$\frac{5}{6}$	$\frac{6}{7}$
Fraction that's been eaten	■	■	■	■	■

4. Which is more, $\frac{9}{10}$ of a pie or $\frac{11}{12}$ of it? How do you know?

Using Benchmarks to Order Fractions

You will need
• Fraction Strips
  (blackline master)

GOAL

Order fractions on a number line.

Cole was playing miniature golf with his friends.
On his first swing, he hit the ball $\frac{5}{8}$ of the way to the hole.

The chart shows how each player did on the first swing.

Cole	Aneela	Joshua	Kate
$\frac{5}{8}$ of the way to the hole	$\frac{4}{5}$ of the way to the hole	$\frac{5}{6}$ of the way to the hole	$\frac{2}{8}$ of the way to the hole

 How can you put the fractions in order from shortest distance to longest?

Cole's Model

I'll use $\frac{1}{2}$ as a **benchmark**.

I can use fraction strips to see which balls went more or less than halfway to the hole.

$\frac{1}{2}$	$\frac{1}{2}$

$\frac{1}{8}$	$\frac{1}{8}$	$\frac{1}{8}$	$\frac{1}{8}$	$\frac{1}{8}$	$\frac{1}{8}$	$\frac{1}{8}$	$\frac{1}{8}$

$\frac{1}{8}$	$\frac{1}{8}$	$\frac{1}{8}$	$\frac{1}{8}$	$\frac{1}{8}$	$\frac{1}{8}$	$\frac{1}{8}$	$\frac{1}{8}$

$\frac{1}{5}$	$\frac{1}{5}$	$\frac{1}{5}$	$\frac{1}{5}$	$\frac{1}{5}$

$\frac{1}{6}$	$\frac{1}{6}$	$\frac{1}{6}$	$\frac{1}{6}$	$\frac{1}{6}$	$\frac{1}{6}$

I know Kate's ball went less than halfway because $\frac{2}{8}$ is close to the starting spot, which is 0.

Aneela, Joshua, and I hit our golf balls more than halfway.

$\frac{1}{2}$				$\frac{1}{2}$			
$\frac{1}{8}$	$\frac{1}{8}$	$\frac{1}{8}$	$\frac{1}{8}$	$\frac{1}{8}$	$\frac{1}{8}$	$\frac{1}{8}$	$\frac{1}{8}$

A. How could you have predicted from the chart that Joshua's ball went farther than Cole's?

B. How could you have predicted that Cole's ball went farther than Kate's?

C. How do you know that Joshua's ball went almost all the way to the hole?

D. Draw a fraction number line to represent the distance the ball has to travel. Use fraction strips to place each golf ball distance on the line.

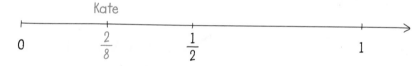

E. Put all the fractions in order from least to greatest.

Reflecting

F. If Jade hits her ball $\frac{3}{5}$ of the way to the hole, is her ball closer to $\frac{1}{2}$ of the way or all the way to the hole? How do you know?

G. If you don't have fraction strips, how can you tell if a fraction is closest to 0, to $\frac{1}{2}$, or to 1?

Hole 2

Checking

1. On the first swings at hole 2, the 4 balls rolled this far along the distance to the hole.

$$\frac{1}{5} \qquad \frac{7}{8} \qquad \frac{6}{10} \qquad \frac{4}{10}$$

   a) Which balls rolled more than halfway to the hole?

   b) Order the fractions from least to greatest.

Practising

2. a) Which fractions are closer to 1 than to $\frac{1}{2}$? Use your fraction strips.

$$\frac{1}{8} \qquad \frac{9}{10} \qquad \frac{3}{8} \qquad \frac{7}{12} \qquad \frac{1}{5}$$

   b) Estimate to mark each fraction on a number line like this.

   0           $\frac{1}{2}$ 1

   c) Order the fractions from least to greatest.

3. Find 3 fractions that fit each description.

   a) between 0 and $\frac{1}{2}$

   b) between $\frac{1}{2}$ and 1

4. Draw a number line from 0 to 1 and use estimation to mark $\frac{2}{10}$. Explain how you knew where to mark the fraction.

5. Write each statement with a denominator that makes it true.

   a) $\frac{2}{\blacksquare}$ is less than $\frac{1}{2}$.

   b) $\frac{3}{\blacksquare}$ is between $\frac{1}{2}$ and 1.

Pot of Gold

Number of players: 3
How to play: Compare fraction cards to see who has the greatest fraction.

You will need
- counters
- a container
- Pot of Gold Game Cards (blackline master)
- a paper bag
- Fraction Strips (blackline master)

- **Step 1** Put the counters in the container. This is the "Pot of Gold." Shake the game cards in a paper bag.

- **Step 2** Each player draws one fraction card. Use fraction strips to figure out whose fraction is the greatest. This player takes one counter. If there's a tie, take another card and compare to see who wins.

- **Step 3** Keep playing until the game card bag is empty.

The player with the most counters wins the game.

Olivia's Comparison

$\dfrac{2}{3}$	$\dfrac{3}{4}$	$\dfrac{1}{12}$
Olivia	Jade	Cole

I know my $\frac{2}{3}$ is greater than Cole's $\frac{1}{12}$ because $\frac{2}{3}$ is almost 1 whole, but $\frac{1}{12}$ is close to 0. I'll use fraction strips to see if my $\frac{2}{3}$ can beat Jade's $\frac{3}{4}$.

Solving Problems by Drawing Diagrams

You will need
- pencil crayons
- Fraction Circles (blackline master)

GOAL

Draw a diagram to solve a problem.

A pizza has toppings on every slice. $\frac{6}{8}$ of the slices have sausage, $\frac{5}{8}$ have peppers, and $\frac{2}{8}$ have mushrooms.

 How many slices could have all 3 toppings?

Joshua's Solution

Understand

The pizza shows 8 equal pieces, or eighths.

There are no slices without toppings.

6 slices have sausage, 5 have peppers, and 2 have mushrooms.

Make a Plan

I'll draw a pizza divided into eighths and use letters to show the toppings.

Carry Out the Plan

I'll start with the sausage because it's on the most slices.

$\frac{6}{8}$ of the pizza means 6 slices.

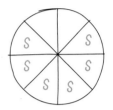

$\frac{5}{8}$ have peppers. That means 5 slices.
I can't leave any empty slices, so
I'll put $\frac{2}{8}$ on the empty slices and
$\frac{3}{8}$ on the other slices.

$\frac{2}{8}$ have mushrooms. That means 2 slices.
There could be 0, 1, or 2 slices with 3 toppings.

0 slices with   1 slice with   2 slices with
3 toppings   3 toppings   3 toppings

Reflecting

A. How did Joshua's diagrams help him solve the problem?

Checking

1. $\frac{3}{5}$ of a group of 5 children are girls, $\frac{4}{5}$ of the group have dark hair, and $\frac{2}{5}$ are wearing jackets. How many children in the group could be dark-haired girls wearing jackets?

Practising

2. This shape represents $\frac{1}{4}$ of a sandbox.
   What could the whole sandbox look like? Show as many possibilities as you can.

3. Jade spent $\frac{3}{10}$ of her allowance to go to a movie. Then she spent $\frac{4}{10}$ to buy a T-shirt for her brother. What fraction of her allowance did she have left?

4. Adam went paddling. After a while, he came back to shore and gave the boat to Kendra. Kendra paddled for twice as long as Adam paddled. If the boat was out for 3 hours altogether, how long did Kendra paddle?

5. When you draw 2 straight lines through a circle, they can cross 0 times or 1 time inside the circle.

0 crossings   1 crossing

How many crossings can there be if you draw 4 straight lines through a circle?

6. Create a problem that someone else could solve by drawing a diagram.

226

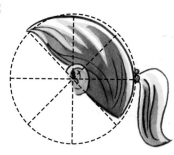

CURIOUS MATH

Drawing with Fractions

You will need
• books about art

Artists often use fraction shapes to help them draw. To draw a side view of a face, you can use a circle divided into eighths.

Step 1

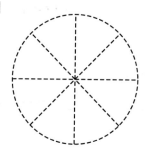

Step 2

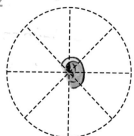

Step 3

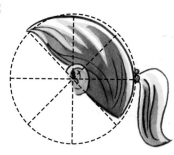

Step 4

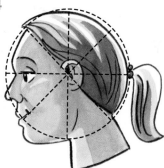

1. How does the circle show that $\frac{8}{8}$ is the same as 1 whole?

2. Use a fraction circle divided into eighths to draw a face picture of your own. Describe how the fraction parts helped you.

3. Look at some pictures other artists have created. How do you think they used fractions to help them draw their pictures?

Mid-Chapter Review

Frequently Asked Questions

Q: How can you represent a fraction?

A: You can colour parts of one whole thing or parts of a group of things. For example, colour $\frac{3}{5}$.

3 of 5 parts

3 of 5 things

Q: How can you compare fractions of the same whole?

 $\frac{4}{5}$

 $\frac{2}{5}$

A1: If the denominators are the same, think about how many parts there are. For example, $\frac{4}{5}$ is more than $\frac{2}{5}$ because the parts are the same size and 4 parts is more than 2 parts.

 $\frac{3}{5}$

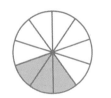 $\frac{3}{10}$

A2: If the numerators are the same, think of the size of the parts. For example, $\frac{3}{5}$ and $\frac{3}{10}$ both describe 3 parts of a whole. Tenths are smaller than fifths, so $\frac{3}{10}$ is less than $\frac{3}{5}$.

Q: How can you decide if a fraction is greater than $\frac{1}{2}$?

A: You can use fraction strips and compare with the $\frac{1}{2}$ strip. For example, $\frac{2}{3}$ is more than $\frac{1}{2}$.

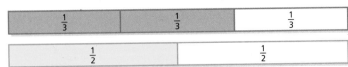

NEL

Practice

1. What 2 fractions does each picture show?

   a)
   b)

2. Colour a picture to show each fraction.

   a) $\dfrac{5}{6}$
   b) $\dfrac{2}{3}$

Lesson 2

3. a) Write 3 fractions that describe the hats.

   b) Put your fractions from part (a) in order from least to greatest.
   c) Sketch a group of hats so that $\dfrac{2}{3}$ are black.

4. Model each fraction with coloured counters. Sketch and colour your model.

   a) $\dfrac{3}{5}$
   b) $\dfrac{6}{10}$
   c) $\dfrac{3}{4}$
   d) $\dfrac{1}{3}$

Lesson 4

5. Put the fractions in order from least to greatest. Explain how you did it.

   $$\dfrac{1}{5} \qquad \dfrac{1}{6} \qquad \dfrac{1}{3} \qquad \dfrac{1}{10}$$

Lesson 5

6. What is each fraction closest to: 0, $\dfrac{1}{2}$, or 1?

   a) $\dfrac{5}{8}$
   b) $\dfrac{2}{10}$
   c) $\dfrac{7}{8}$
   d) $\dfrac{1}{6}$

Lesson 7

Decimal Tenths

You will need
- counters

GOAL

Represent fraction tenths as decimals.

Tien has rings on all of her fingers.

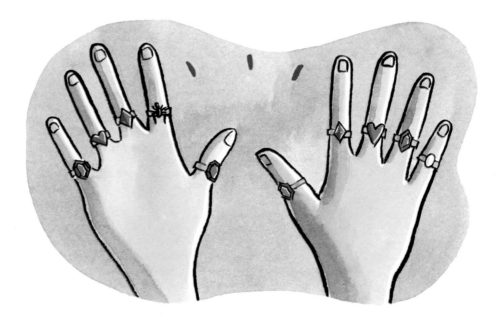

❓ **How can Tien describe the number of blue rings?**

Tien's Fractions

I have 2 blue rings.
That's *two tenths* or *zero and two tenths* of my rings.
I can write the fraction $\frac{2}{10}$.

NEL

decimal

A way to describe fractions using place value. A decimal point separates the ones place from the tenths place.

I can also write two tenths as a **decimal**.

Zero means there is less than one whole.

0.2 ← number of tenths

↑
decimal point

I can show 0.2 on a fraction strip.
There are 10 rings, so I'll use a tenths strip.

◈	◈								

0.1 0.2

$\frac{1}{10}$ $\frac{2}{10}$

Reflecting

A. Why are rings on fingers a good model for decimal tenths?

B. How do these counters show that 0.2 of the rings are blue?

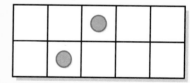

C. 1.0 means $\frac{10}{10}$ or 1 whole group. Why could Tien say that 1.0 tells about the rings that have gold bands?

Checking

1. Two tenths or $\frac{2}{10}$ or 0.2 of Tien's rings are blue. Show more ways to describe Tien's rings with fractions and decimals.

Practising

2. Write each as a fraction and as a decimal.
   a) three tenths
   b) ten tenths
   c) one tenth
   d) zero tenths

3. Use counters on the grid to show each decimal.
   Sketch each model.
   a) 0.3   b) 0.7

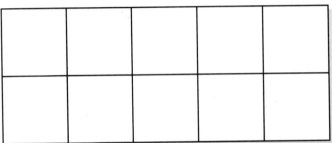

4. What comes next in this counting pattern? How
   do you know?

   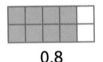

   0.2   0.4   0.6   0.8

5. Which colours and pictures show 0.3?
   A. 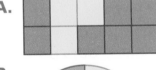   C

   B. 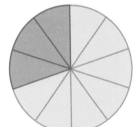   D.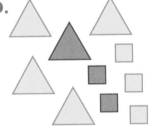

6. a) Trace your 2 hands and draw rings on your
      fingers. Use at least 3 different colours.
   b) Use words and decimals to describe your rings.

7. How are pictures for 0.8 and 0.2 alike? How are
   they different?

232

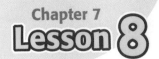

Lesson 8

Decimal Hundredths

You will need
- a 100 grid
- pencil crayons

GOAL

Represent fraction hundredths as decimals.

 How can you use decimals to describe the colours in a picture?

Ethan's Picture

My grid is divided into 100 parts.
1 part or *one hundredth* of the grid is yellow.
I write the fraction $\frac{1}{100}$ as the decimal 0.01.

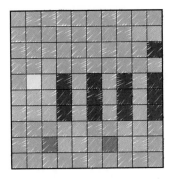

A. Ethan's whole grid is represented by $\frac{100}{100}$ or 1.00 or 1. Which colour covers $\frac{28}{100}$ or 0.28 of the grid?

B. Which colour covers $\frac{10}{100}$ or 0.10 or 0.1 of the grid?

C. Write a fraction and a decimal in hundredths for every other colour in Ethan's picture.

Reflecting

D. How do you know that 1.00 is the same amount as 1.0 or 1?

E. How is 0.07 different from 0.70? Use a 100 grid to explain.

Checking

1. Write a fraction and a decimal for the green part.

a)

b)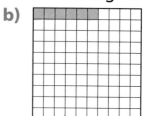

Practising

2. Write a fraction and a decimal for the yellow part.

a)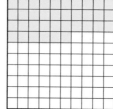

c)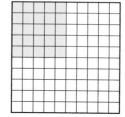

b)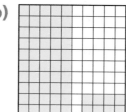

d)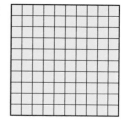

3. Write each fraction as a decimal.

a) $\dfrac{89}{100}$

c) 30 hundredths

b) $\dfrac{9}{100}$

d) 12 hundredths

4. Write each decimal as a fraction.

a) 0.67   b) 0.29   c) 0.40   d) 0.04

5. a) How many centimetres would you use to measure a line that's 0.31 m long?

Metre stick (100 cm)

b) What are 2 different decimal names for one whole metre stick? Explain how you know.

6. Look at the statements about the 100 chart below.
Decide whether each statement is true or false.
If the statement is false, rewrite it to make it true.

1	2	3	4	5	6	7	8	9	10
11	12	13	14	15	16	17	18	19	20
21	22	23	24	25	26	27	28	29	30
31	32	33	34	35	36	37	38	39	40
41	42	43	44	45	46	47	48	49	50
51	52	53	54	55	56	57	58	59	60
61	62	63	64	65	66	67	68	69	70
71	72	73	74	75	76	77	78	79	80
81	82	83	84	85	86	87	88	89	90
91	92	93	94	95	96	97	98	99	100

a) On a 100 chart, 0.10 of the numbers have a 0 in them.
b) On a 100 chart, 0.17 of the numbers have a 2 in them.
c) On a 100 chart, 0.1 of the numbers have 3 digits.
d) On a 100 chart, 0.58 of the numbers are greater than 42.

7. Write 2 more true statements about numbers on a 100 chart. Use decimals in your statements.

8. a) Create your own design on a 100 grid. Use at least 4 colours.
   b) Use fractions and decimals to describe the colours in your design.
   c) Which colour is closest to $\frac{1}{2}$ of your design? Explain how you know.

9. Describe an everyday situation in which someone might use decimal hundredths.

Chapter 7

Lesson 9

Representing Decimals with Coins

You will need
- a 100 grid
- pennies and dimes

GOAL

Represent parts of a dollar as decimals.

? **How can you pay for this toy with pennies and dimes?**

Julia's Way

I can think of 1 dollar as a 100 grid.
There are 100 squares in my grid.
There are 100 cents in a dollar.

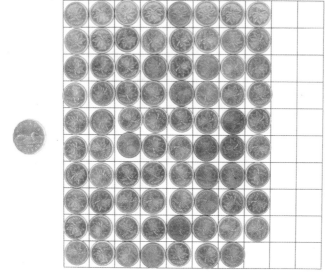

Each cent is worth $\frac{1}{100}$ or 0.01 of a dollar.

$0.79 means $\frac{79}{100}$ of 1 dollar, or 79¢.

I can pay with 79 pennies.

236

Luis's Way

A column of pennies on a 100 grid is worth a dime, or $\frac{10}{100}$ or 0.10 of a dollar. Since there are 10 dimes in a dollar, a dime is also worth $\frac{1}{10}$ or 0.1 of a dollar.

Julia made $0.79 on a 100 grid with 7 columns of pennies and 9 more pennies.

Each column is worth a dime, so I can also pay with 7 dimes and 9 pennies.

Reflecting

A. How much are 7 columns of pennies on a 100 grid worth?

B. You can read 0.79 as *seventy-nine hundredths* or *seven tenths and nine hundredths*. What does this have to do with pennies and dimes?

C. How can you use grids and pennies to show $1.79?

Checking

1. Julia pays with pennies. Luis pays with pennies, dimes, and loonies. Show how each person would pay each amount.
   a) $0.45 **b)** $0.92 **c)** $2.08 **d)** $1.20

2. How can you represent each amount of money as a decimal?

a)  b)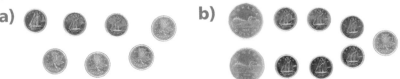

Practising

3. How would Julia and Luis pay each amount?
   a) $0.24   b) $0.03   c) $0.92   d) $0.40

4. How can you represent each amount of money as a decimal?

a)

d)

b)

e)

c)

f)

5. How much money does each 4 in $4.44 represent?

6. You have 9 coins, which can be pennies or dimes.
   a) What is the greatest amount of money you could have? Write it as a decimal.
   b) What is the least amount of money you could have? Write it as a decimal.

7. Why does it make sense that you can represent cents as hundredths?

Race to 1

Number of players: 2 to 4

How to play: Spin and colour hundredths on a grid.

You will need
- Fraction Circles (blackline master)
- a paper clip and a pencil (for spinner)
- a 100 grid for each player
- pencil crayons

- **Step 1** Make a spinner like this one using a fraction circle.

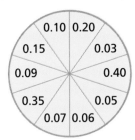

- **Step 2** Take turns spinning. Colour the numbers you spin on your 100 grid.

Continue playing until someone has coloured one whole grid.

Lang's Turn

I had 90 hundreds coloured. Then I spun 0.03, so I coloured 3 hundredths. If I spin 0.07 or more this time, I can win!

Estimating Decimal Sums and Differences

GOAL

Estimate sums and differences with decimal tenths and hundredths.

The Kids Guide to Money Cents by Keltie Thomas says that keeping a money diary can help you save money.

Kate is saving to buy a skateboard.
Her weekly allowance is $15.00.
She spends some and puts the rest in a savings account.

Kate's Money Diary

Sun.	Mon.	Tues.	Wed.	Thurs.	Fri.	Sat.
None	None	None	Forgot water for gymnastics. Bought one from machine. Spent $1.75.	Went to store. Spent $2.39 for a snack and a drink.	None	Went to fun fair at the park. Spent $5.85. Put the rest of my money in the bank.

 About how much money did Kate put in the bank this week?

Cory's Estimate

First I need to know about how much Kate spent.
Then I'll subtract the total from $15.00.
I don't need to know the exact amounts, so I'll use whole-dollar amounts that are close.

On Wednesday, she spent $1.75.
That's about $2.00.

On Thursday, she also spent about $2.00.

I'll add these, then I'll add an amount for Saturday.

Hailey's Estimate

I'll estimate a different way.
Kate spent about $1.70 on Wednesday and about $2.30 on Thursday. She spent more on Saturday.

A. About how much money did Kate spend this week? Show the numbers you used to estimate.

B. About how much money did Kate have left to put in the bank?

Reflecting

C. Why didn't you need to use the exact amounts to solve the problem?

D. Why do you think Hailey used $1.70 and $2.30 for her estimate?

What I Spent
This Week
Monday: $7.45
Wednesday: $6.25
Saturday: $11.85

Penny's Pitas

	Yogurt	$ 1.29
	Pita	$ 2.75
	Juice	$ 1.25
	Milk	$ 0.89
	Soup	$ 1.39
	Fruit Salad	$ 1.85

Checking

1. About how much money did Ethan spend this week? Show the numbers you used to estimate.

2. If Ethan had $26.89 to start with, about how much money does he have left?

Practising

3. Jade ordered a pita and soup. Gen ordered yogurt.
   a) Estimate the cost of Jade's order.
   b) About how much more money did Jade spend than Gen? Explain how you know.

4. Michael has $3.50. He wants soup, milk, and a pita. Does he have enough money? Explain how you know.

5. Choose 2 items from the Penny's Pitas menu.
   a) Estimate the difference in cost.
   b) Estimate the total cost of both items.

6. Create your own estimating problem about the menu. Show how to solve your problem.

7. Julia estimated $3.49 + $19.58 to be about $23. Do you agree? Explain.

8. Describe 2 different ways to estimate $12.70 − $8.49.

Using Mental Math

> **GOAL**
>
> Add and subtract decimals using mental math strategies.

The principal is retiring.
Luis's class raised $50.00 for a gift.
They are going to buy a plant and a wall decoration.

 How much money will the class have left from $50.00?

Luis's Method

I'll add 1¢ to each price to make it easier to add in my head.

$25.00 + $15.00 = $40.00

A. What do you think Luis will do next to calculate the actual cost?

B. How much money will be left? Show your work.

Reflecting

C. Why does it make sense to use mental math to calculate the cost of the gifts?

D. How can you count up to check your answer for Part B?

Checking

1. You have $50.00 to spend at this sale.

a) Choose 2 items you could buy. Use mental math to calculate the total cost. Write the numbers you used.

b) Use mental math to calculate the amount you will have left. Write the numbers you used.

Practising

2. Kate had $14.11 in her wallet in the morning. She had $9.15 left that night. Now she's wondering how much money she spent. See the photo below.

a) Why do you think Kate says this?

b) How could you use Kate's idea to figure out how much money is left?

I wish I had $9.11 left instead of $9.15. Then it would be easy to figure out!

3. Adrian spent $9.95 on a shirt and $1.49 on socks. Use mental math to figure out how much he spent altogether. Write the numbers you used.

4. This speech bubble shows a mistake people sometimes make when they use mental math to subtract decimals. Write a new speech bubble to show a correct way to figure out how much money is left.

> I had $25.00 and I spent $22.75.
>
> $25.00 − $22.00 = $3.00, and
>
> $1.00 − 75¢ is 25¢, so I should have
>
> $3.25 left.

5. Use mental math to add or subtract. Write the numbers you used.
   a) $10.99 + $6.99
   b) $42.00 − $31.80
   c) $14.49 + $29.51
   d) $23.25 − $15.75

6. Ethan won a $25.00 gift certificate in a contest. He used it to buy a CD that cost $14.95 and a poster that cost $5.99.
   a) Use mental math to figure out how much Ethan has spent so far.
   b) Use mental math to figure out how much money he has left.

7. Which of these would you calculate using mental math? Why?
   A. $15.00 − $11.73
   B. $14.27 − $11.73

The MUSIC STORE $25.00
Gift certificate

Lesson 12 Making Change

You will need

- play money coins and bills

GOAL

Count change for a given purchase.

Ethan has a $20 bill. He plans to buy the skateboard at the yard sale.

② How much change will Ethan get?

Ethan's Solution

My change should be less than $2.00 because
$18.00 + $2.00 = $20.00.

I can count on from $18.35 to $20.00 to figure out the actual amount.

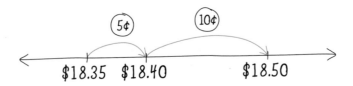

A. How might Ethan continue counting? Use money and a number line.

B. How much change will Ethan get from $20.00? Use your number line and counting back.

Reflecting

C. Draw a different number line to count Ethan's change. Start with a $1.00 jump.

D. Show another way to count Ethan's change. Use coins and bills to show what you did.

E. Ethan used counting on to count the change. Why might he think of this as adding? Why might he think of this as subtracting?

Checking

1. Darra wants to use  to buy .

   a) Estimate the amount of change she will get.
   b) Calculate the amount of change she will get.

Practising

2. Estimate and then calculate the change.

   a)   b)

3. You have  to spend at the yard sale.

   Choose an item you could buy. Which bill or bills would you use to pay for the item? Show how to calculate your change.

4. Describe 2 ways to count change. Give an example for each way.

Adding and Subtracting Decimals

GOAL

Add and subtract decimal tenths and hundredths using number lines.

A baby boa constrictor was about 0.65 m long. It grew up to be 1.97 m long.

 **How much did the boa constrictor grow?**

Emily's Solution

I'll estimate first.

I know that 1.97 m means 1 whole metre plus 0.97 of another metre. That's almost 2 m!

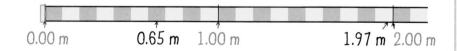

0.00 m   0.65 m   1.00 m   1.97 m 2.00 m

The number line can help me estimate.

I know that 0.65 m is close to 0.5 m, and 1.97 m is close to 2.0 m.

If the snake grew from 0.5 m to 2.0 m, then that would be 1 whole metre and another 0.5 m.
The snake grew about 1.5 m.

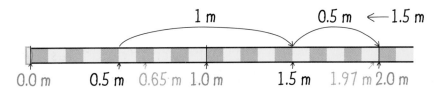

I can also use the number line to count on from 0.65 m to 1.97 m.

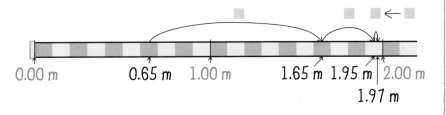

A. Emily drew 3 jumps to figure out how far it was from 0.65 m to 1.97 m. How long was each jump?

B. How much did the boa constrictor grow?

Reflecting

C. How did the second number line help Emily make sure her answer made sense?

D. How did the third number line help Emily calculate how much the snake grew?

Checking

1. A baby bull snake grew from 0.42 m to 1.53 m.
   a) Use tenths to estimate how much the snake grew.
   b) Use a number line to calculate how much the snake grew. Show how you figured it out.

2. Another baby snake was 0.45 m long. It grew 0.80 m. How long did it grow to be? Estimate first. Then use a number line.

Practising

3. A giraffe's neck can be about 2.5 m long. If a giraffe is about 5.4 m tall, how tall is the rest of its body? Show your work.

4. How much more does an adult ticket cost than a child ticket? Estimate to predict or check your answer.

a)
Tickets	
Adults	$ 5.50
Children	$ 1.49

b)
Tickets	
Adults	$ 3.25
Children	$ 1.89

5. Add these masses. Estimate to predict or check.
   a) 6.2 kg and 1.3 kg   b) 9.21 kg and 3.25 kg

6. Emily says that 3.4 m + 2.8 m is a good way to estimate 3.49 m + 2.82 m. Do you agree? Explain.

7. How could Alec finish each part of his journal entry?

> Adding and subtracting decimals is just like adding and subtracting whole numbers because ...
>
> The only difference is that...

Chapter Review

Frequently Asked Questions

Q: What are decimal tenths and hundredths?

A: A decimal is part of a whole. For example, 0.8 means $\frac{8}{10}$ of a whole. It is just another way to write a fraction. Each picture below shows 0.8, or 8 tenths.

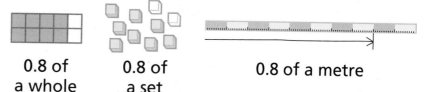

0.8 of
a whole

0.8 of
a set

0.8 of a metre

Decimals can also show hundredths. For example, each picture below shows 0.12 or $\frac{12}{100}$. The money picture shows that 0.12 is also one tenth and two hundredths.

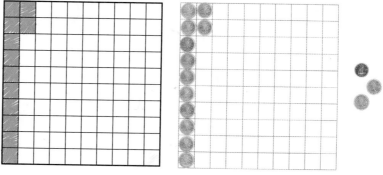

0.12 of a whole

2 ways to show 0.12
of a dollar

Q: How can I add or subtract decimal numbers?

A1: Decide if you need to know the exact answer or if you can estimate. To estimate, think of whole numbers that are close to the decimals, or look for decimals that fit together to make estimating easier. For example, to estimate 10.82 + 3.14, you might say, "It's close to 11 + 3, so the answer is close to 14." You might also say, "It's close to 10.8 + 3.2. I can add the 0.8 and 0.2 to get one whole, so that's 10 + 3 + 1 = 14."

A2: If you need the exact answer, it can help to think of money. For example, 2.90 + 3.99 is almost 3.00 + 4.00, but you have to subtract a dime and a penny.

> $3.00 and $4.00 is $7.00.
>
> $7.00 minus 10¢ is $6.90,
>
> and $6.90 minus 1¢ is $6.89.

A3: Another way to add or subtract decimal numbers is to draw a number line and draw jumps that are easy to make using mental math. For example, if you have to subtract 6.5 − 3.4, you could draw this number line and count up from 3.4 to 6.5 to figure out the difference.

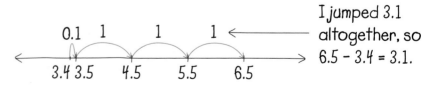

I jumped 3.1 altogether, so 6.5 − 3.4 = 3.1.

Practice

Lesson 1

1. Write the fraction for the coloured part of each shape.

   a)    b)   c)

2. Which shape is $\frac{2}{3}$ yellow?

   A.   B.    C.

Lesson 2

3. Write 3 sentences to describe this group of shapes. Use a different fraction in each sentence.

4. Sketch a group of shapes that shows the fraction $\frac{5}{8}$.

Lessons 4 and 5

5. Put these fractions in order from least to greatest.

   $\frac{2}{5}$ $\frac{2}{12}$ $\frac{2}{6}$ $\frac{2}{3}$

6. Draw a number line from 0 to 1. Label $\frac{1}{2}$. Use estimation to place each fraction on the line.

   $\frac{4}{10}$ $\frac{3}{10}$ $\frac{1}{12}$ $\frac{10}{12}$

Lesson 6

7. $\frac{4}{5}$ of the 5 basketball team members are very tall.

   $\frac{2}{5}$ of the team members are excellent dribblers. How many of the tall players could be excellent dribblers? Is there more than one answer? Explain how you know.

Lessons 7 and 8

8. Write a decimal and a fraction for each coloured part.

a) b)

Lesson 9

9. How can you represent each amount of money as a decimal?

a)  b)

Lesson 10

10. Estimate. Write the numbers you used.
    a) $3.85 + $4.98 + $2.15   b) 14.35 − 8.79

Lesson 11

11. Use mental math. Write the numbers you used.
    a) $3.99 + $5.99   b) $22.00 − $6.49

Lesson 12

12. Julia had $9.00 to spend at a yard sale.
    She bought a hockey stick for $7.75.
    How much change did she get?

Lesson 13

13. An African elephant was 0.90 m tall at birth.
    It grew up to be 3.17 m tall.
    How much did it grow?

What Do You Think Now?

Look back at **What Do You Think?** on page 207. How have your answers and explanations changed?

Chapter Task

Fraction Kites

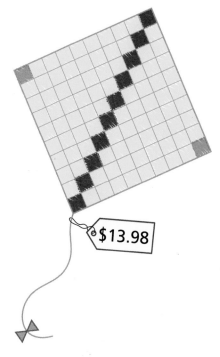

$13.98

Task Checklist

✔ Do your kites fit the rules?

✔ Did you use fractions to describe all your kites?

✔ Did you use decimals if you could?

✔ Did you estimate the costs?

✔ Did you explain your answers to Parts D and E?

Jade sent her idea for a fraction kite to the kite company. They loved it! They want Jade to design 3 more kites.

- Each kite is divided into a different number of equal parts.
- One kite shows tenths.
- Each kite has at least 3 colours.
- One kite is less than half yellow.
- The new kites will cost $9.95, $10.98, and $18.89.

❓ What might the kites look like?

A. Use fractions and decimals to describe Jade's kite.

B. Design 3 new kites that fit the rules. Label each kite with one of the 3 prices above.

C. Use fractions and decimals to describe your kites.

D. Which 2 kites would cost a total of about $30?

E. About how much more does the most expensive kite cost than the least expensive one?

Ticket sales

K

Gr. 1 🐟 🐟 🐟

Gr. 2 🐟 🐟

1. The pictograph shows the number of students who bought tickets for the fish pond at the penny carnival. The students bought 45 tickets altogether.

   How many tickets does each 🐟 represent?

   A. 2   **B.** 3   **C.** 4   **D.** 5

2. How many tickets did the Grade 1 class buy?

   A. 3   **B.** 9   **C.** 15   **D.** 30

3. Where would 598 and 499 go in the Venn diagram?
   A. 598 with red numbers; 499 with blue numbers
   B. 598 with red numbers; 499 with green numbers
   C. 598 with blue numbers; 499 with red numbers
   D. 598 with red numbers; 499 with red numbers

Numbers

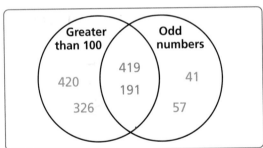

4. Which shape is not symmetrical?

   A.

   C.

   B.

   D.

5. What number does ▪ represent in this equation?

$9 \times 8 = $ ▪

A. 74 B. 17 C. 72 D. 68

6. Which choice does *not* describe this array?
   A. $42 - 7 - 7 - 7 - 7 - 7 - 7$
   B. $6 + 6 + 6 + 6 + 6 + 6$
   C. 6×7
   D. $42 \div 7$

7. What number does ▪ represent in this equation?

$54 \div 6 = $ ▪

A. 7 B. 8 C. 9 D. 48

8. Which calculation has the greatest quotient?
   A. $32 \div 4$ B. $32 \div 8$ C. $14 \div 2$ D. $6 \div 1$

9. Which fraction is closest to $\frac{1}{2}$?

A. $\frac{5}{9}$ B. $\frac{2}{9}$ C. $\frac{9}{9}$ D. $\frac{7}{9}$

10. Which set of fractions is in order from least to greatest?

A. $\frac{4}{5}, \frac{4}{10}, \frac{4}{8}$

C. $\frac{4}{10}, \frac{4}{8}, \frac{4}{5}$

B. $\frac{4}{4}, \frac{4}{8}, \frac{4}{10}$

D. $\frac{4}{4}, \frac{4}{10}, \frac{4}{5}$

11. Which decimal is equal to $\frac{8}{100}$?

A. 0.88 B. 8.1 C. 0.80 D. 0.08

12. Which number does *not* mean 0.44?
    A. forty-four tenths
    B. forty-four hundredths
    C. zero ones and forty-four hundredths
    D. zero ones, 4 tenths, and 4 hundredths

13. Which number is two tenths less than 1.1?
    A. 1.12 B. 1.18 C. 0.9 D. 0.8

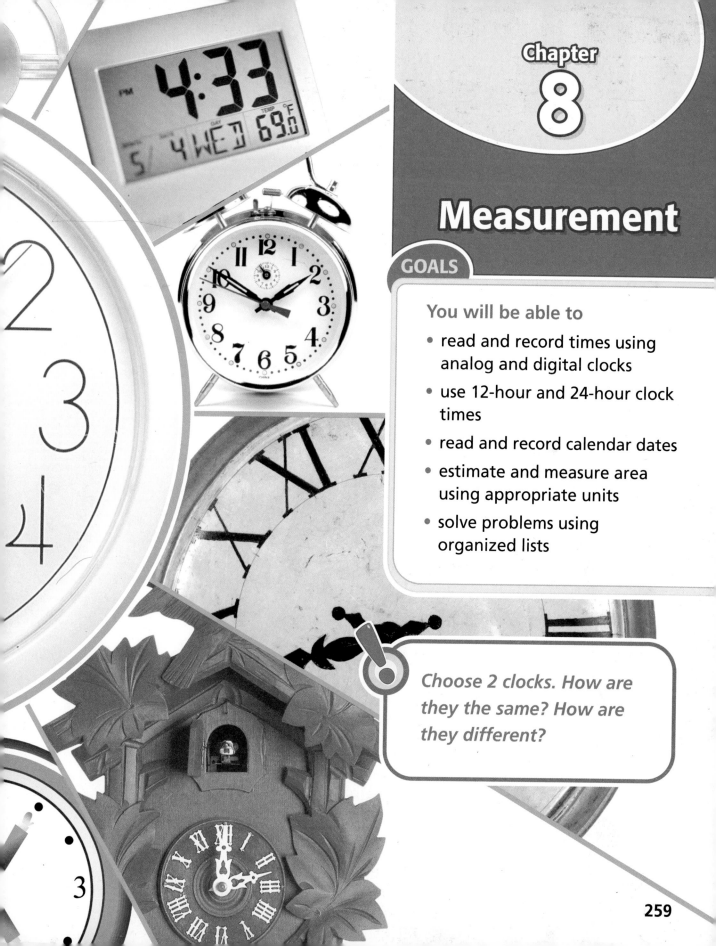

Measurement

GOALS

You will be able to

- read and record times using analog and digital clocks
- use 12-hour and 24-hour clock times
- read and record calendar dates
- estimate and measure area using appropriate units
- solve problems using organized lists

Choose 2 clocks. How are they the same? How are they different?

Getting Started

- a way to measure 1 minute

Measuring Time

Jade can do 67 jumping jacks in 1 minute.

NEL

? What can you do in 1 minute?

A. How many jumping jacks can you do in 1 minute? Explain how you found out.

B. What other exercises can you do in 1 minute? Use a clock or a timer to help you find at least 2 more 1-minute exercises.

What Do You Think?

Do you *agree* or *disagree* with each statement? Explain your thinking.

1. I would use minutes to measure the amount of time I spend sleeping in 1 week.

2. I can count past 1000 in 1 hour.

3. A student born in March must be older than a student born in June.

4. There is only one way to write today's date.

Lesson 1 — Telling Time to the Hour

You will need
- Timelines (blackline master)
- Analog Clocks (blackline master)

GOAL

Read and record clock times to the hour, and identify activities for different times.

A timeline shows all the hours in a day.
You can use a timeline to help you plan your day.

 How can you use times to show your plans for tomorrow?

a.m.

From midnight to before noon

p.m.

From noon to before midnight

Ethan's Timeline

I'll start by marking when I get up and when I go to bed. I always get up at *7 o'clock in the morning.*

That's 7:00 **a.m.**

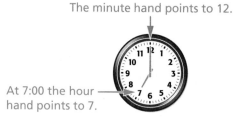

The minute hand points to 12.

At 7:00 the hour hand points to 7.

Analog clock

Digital clock

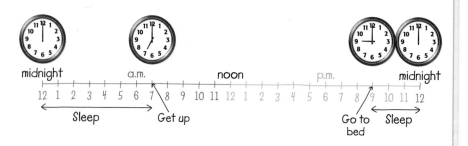

midnight   a.m.   noon   p.m.   midnight

12 1 2 3 4 5 6 7 8 9 10 11 12 1 2 3 4 5 6 7 8 9 10 11 12

← Sleep → Get up   Go to bed ← Sleep →

A. At what time does Ethan go to bed?
Write the time 2 ways with numbers and words.

B. Make a timeline that shows the hours from
midnight tonight until midnight tomorrow. Use
arrows and labels to show your plans for the day.
Show at least one thing for a.m. and one for p.m.

Reflecting

C. How did you make sure your timeline
showed the whole day?

D. Pick one of the arrows on your timeline.
What would analog and digital clocks look
like for that time?

Checking

1. Carmen's guitar lesson starts
at this time after school. Write
the time 2 ways with numbers
and words.

Practising

2. Write these *morning* times 2 ways with numbers
and words.

a)

b)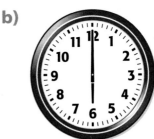

3. Write these times using numbers and a.m. or p.m.
   a) 4 o'clock in the afternoon
   b) 8 o'clock at night

4. Hélène is at school from 9:00 a.m. until 4:00 p.m. How many hours is that? How do you know?

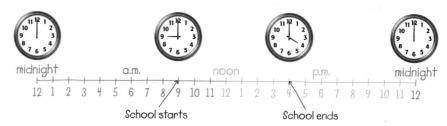

5. What will you likely be doing tomorrow at each time?
   a) 9:00 a.m.   b) 7:00 p.m.

6. How are these times alike?
   How are they different?

7. If you went to bed at 9:00 p.m. and got up at 7:00 a.m. the next day, how many hours of sleep would you get? How do you know?

8. Why do we need a.m. and p.m. to describe times on a clock?

9. How does each type of clock show *o'clock* times? Give an example for each.
   a) an analog clock   b) a digital clock

Clock Fractions

You can use fractions to make a clock face.

- **Step 1** Trace something round to make a circle.

- **Step 2** Cut out the circle.

- **Step 3** Fold the circle in half and then in half again.

- **Step 4** Now you have a wedge shape. Fold the wedge in thirds so the 3 parts fit on top of each other. Then unfold.

1. How many equal parts do you see in your circle?

2. Write numbers on the fold lines to make a clock face.

3. What fraction of the circle is between the lines?
   a)
   b)

4. Draw hands on your clock to make an *o'clock* time.
   a) What time did you show?
   b) What fraction did you show?

Time to the Half Hour and Quarter Hour

You will need
- a clock with movable hands

GOAL

Read and record clock times to the half hour and quarter hour.

Kate is baking a cake after dinner.

Start Time	Checking Time	Finish Time

 When did Kate open the oven door?

Kate's Cake Times

For all the times, the hour hand shows it is after 6 o'clock and before 7 o'clock.

Start time: The minute hand points to 3. That's 15 minutes after the hour, or 6:15 p.m. The starting time is *six fifteen*, or *a quarter after six*.

Checking time: The minute hand points to 6. That's 6:30 p.m., or *six thirty*, or *half past six*.

Finish time: The minute hand points to 9. That's 6:45 p.m., or *six forty-five*, or *a quarter to seven*.

60 minutes in 1 hour

00:45 00:15
00:30

Reflecting

A. What is different about the hour hand for Kate's 3 times?

B. Why do we use fraction names like *half past* and *a quarter to* when we talk about times?

Checking

1. Write the time with numbers and with fraction words.

a)    b)    c)

Practising

2. Write the time with numbers and with fraction words.

a)    b)    c)

3. Josef rode his bike from 11:30 a.m. until 11:45 a.m. How long did he ride? Explain what you did.

4. What can you tell about the time?
   a) The minute hand is pointing to 9.
   b) The hour hand is pointing to 9.

Telling Time to 5 Minutes

You will need
* Analog Clocks
  (blackline master)

 GOAL

Read and record clock times to the 5-minute marks.

Cory is chatting online with his friend Andrew in

Alberta. For Cory, in Manitoba, the time is  .

For Andrew, in Alberta, the time is 1 hour earlier.

❓ What time is it in Alberta?

Cory's Time

When I read a clock, I look at the hour hand first. The hour hand shows that it is after 2 o'clock and just before 3 o'clock.
The minute hand is pointing to the 10.

You can skip count by 5s to tell the minutes after 2 o'clock. The time is *fifty minutes after two* (2:50) in Manitoba. Other names for 2:50 are *two fifty* and *ten minutes to three*.

A. In Alberta, the time is 1 hour earlier than 2:50. Use a clock face to show the Alberta time.

B. Write the Alberta time using numbers and using *to* or *after*.

Reflecting

C. Why do you think Cory looks at the hour hand first?

D. How do you know that ten minutes to three is another name for 2:50?

Checking

1. Cory plans to talk to Angie online at this time.
Where Angie lives, the time is 2 hours later than in Manitoba. At what time should Angie go online?

Practising

2. Write the time using numbers and using *to* or *after*.

a) 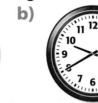 b)  c)

3. Draw each time on a clock face.

    a) 5:20   b) 2:05   c) 1:40

4. Natalie starts soccer practice
   at this time.
   Practice lasts for 2 hours.
   When will Natalie's practice end?

5. Paulette left school
   at this time.

   She got home at
   this time.

   How long did it take her to get home?
   How do you know?

6. The hour hand on a clock is a bit more than
   halfway between 4 and 5. About what time is it?
   How do you know?

7. Why do you think we say *ten to three* but not *fifty
   to three*?

8. How can knowing how to count by 5s help you
   read the time on a clock? Give an example.

Clocks and Locations

In Clock Ball, the person in the centre calls out clock times instead of names to tell where the ball will go next.

Luis visualizes himself at the centre of a clock, facing Ethan at 12 o'clock.

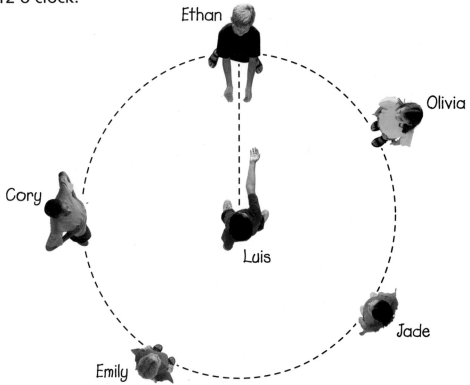

1. a) What hour would Luis call out for Olivia?
   b) Who is standing at 9 o'clock?

2. a) Draw a circle with your name at the centre.
   b) Draw 5 dots at o'clock times around the circle. Label each dot with a name.
   c) Trade drawings with a classmate. Write the clock time that tells where each person is standing.

Lesson 4

Telling Time to 1 Minute

GOAL

Read clock times to the nearest minute on different kinds of clocks.

Luis wants to set the time on his new analog watch. His digital clock shows the correct time.

? **Where should Luis set the hands on his watch?**

Luis's Watch

It's 1:47. That's after 1:00, but before 2:00.

First I'll turn the knob till the hour hand gets to 1. Then I can keep turning to set the minutes.

1:00

1:47 means 47 minutes after 1. That's between 1:45 and 1:50. When I move the minute hand 45 minutes around the dial, the hour hand moves away from 1 and toward 2.

1:45

To get from 1:45 to 1:47, I go 2 more minutes: 1:46, 1:47. The marks between the numbers show minutes. Now my watch is set to *one forty-seven*.

1:47

Another way to say that is *thirteen minutes to two*.

Reflecting

A. How does Luis's watch show that the time is almost 2:00?

B. How does the digital clock show that the time is almost 2:00?

C. How will Luis's watch and digital clock change when one more minute passes?

Checking

1. When the digital clock shows this time, where will the minute hand on Luis's watch be pointing?

2. a) When Luis's watch shows this time, what will his digital clock show?
   b) Write 2 ways to say the watch time using *to* or *after*.

Practising

3. Write the time using numbers.

   a)    b)    c)

4. Write the time using words in a way that tells the number of minutes to the next hour.

   a)    b)    c)

5. Samia painted from 10:45 a.m. to 12:35 p.m. at *L'École des jeunes artistes*.
   How long did she paint? Use drawings or words to show how you used the clock or the timeline to solve the problem.

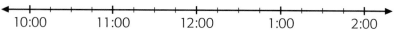

6. A concert begins at 8:00 p.m. and ends at 9:30 p.m. Nicky and Jon look at their watches during the concert. Which watch makes it easier to see how much longer the concert will last? Why?

Nicky   Jon

7. Which watch from Question 6 makes it easier to see how many minutes have passed since the concert began?

8. What time will Luis's digital clock show after 2:59? How do you know?

9. a) Use the digits 2, 4, and 6 to write 4 times you could see on a digital clock. Make 2 of the times a.m. Make the other 2 times p.m.
   b) Put your clock times in order from earliest to latest.
   c) Choose 2 of your clock times. Write a sentence to describe what you might be doing tomorrow at each time.

10. Write your own problem about how much time something took. Show how to use a clock or a timeline to solve your problem.

11. How do you read the time on an analog clock? Give an example.

Writing Dates and Times

GOAL

Write numeric dates and 24-hour times.

numeric date

A way of writing dates with numbers and no words. One way is year-month-day; for example, 2002-08-15.

Diane made cards using **numeric dates** to show when her cousins were born.

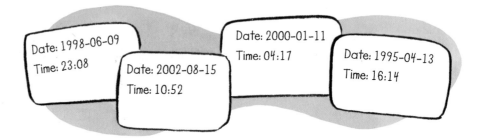

Date: 1998-06-09
Time: 23:08

Date: 2002-08-15
Time: 10:52

Date: 2000-01-11
Time: 04:17

Date: 1995-04-13
Time: 16:14

 If Diane's cousin Heather is the eldest, when was she born?

Diane's Dates and Times

One of my cousins was born on 1998-06-09. I wrote the year first, then the month number, and then the day number.
That's the 9th day of the 6th month of 1998.
In calendar form, I write June 9, 1998.

I used **24-hour times**.
24-hour times don't use a.m. and p.m.
Midnight is 00:00. The last minute before midnight is 23:59.

My cousin was born at 23:08. That means 11:08 p.m.

Communication Tip

On the 24-hour clock, 13:00 means 1:00 in the afternoon.

A. On what date was the youngest cousin born? How do you know?

B. How many cousins were born in the morning? Write their birth times using a.m.

C. On what date was Heather born? How do you know?

D. At what time was Heather born? Write her birth time using 12-hour time and a.m. or p.m.

Reflecting

E. Which 24-hour time do you think is easier to change to a.m. or p.m. form? Explain.

F. Why is it easier to change a 24-hour time to p.m. form when you're using a clock like the one on the left?

G. If 2 students in your class wrote their birth dates in numeric form, what part of the date would you look at first to see who is older? Why?

Checking

1. Rey's birth date is 1999-09-19. His birth time is 19:19.

   a) Write Rey's birth date using calendar form.

   b) Write Rey's birth time using a.m. or p.m.

Practising

2. Write each calendar date using numeric form.
   a) January 12, 1959
   b) March 15, 1989

3. Write each numeric date using calendar form.
   a) 1867-07-01
   b) 2004-10-30

4. Other ways to write 2006-02-18 are

   | 18-02-2006 | 18/02/2006 | 02/18/2006 |

   Write March 15, 2008 in 3 ways using numeric form.

5. Write each time using a.m. or p.m.

   a)
   b)

6. Write each time using 24-hour form.
   a) the time you wake up
   b) the time you go home from school

7. How can visualizing a timeline help you change a 12-hour time to a 24-hour time?

8. Would a 24-hour clock ever show these times? Explain.
   a) 16:63
   b) 03:33
   c) 27:15
   d) 01:01

9. a) If a date is 17/03/2008, what calendar date is it?
   b) How can you tell which number is the month?
   c) How can you write your birth date this way?
   d) If a date is 06/03/04, what could it mean?
   e) Why is it important to make sure people know which method you're using to write numeric dates?

10. What are some advantages and disadvantages of using 24-hour times and numeric dates?

NEL

It's About Time

You will need

- 4 sets of number cards for 0 to 9
- a p.m. and an a.m. card for each player
- a : card for each player

Number of players: 2 or 3

How to play: Arrange cards to show times, and decide which time is later.

- **Step 1** Shuffle the number cards. Deal 4 cards to each player.

- **Step 2** Players arrange 3 or 4 cards to make a 12-hour time or a 24-hour time. The latest possible time is 11:59 p.m.

- **Step 3** If the numbers can't be used to make a time (for example, 7, 8, 9, 9), the player discards 1 card and takes a new one. If a time still can't be made, the player misses a turn.

- **Step 4** The player who makes the latest time gets a point. If 2 players make the same time, both get a point. Keep playing until one player has 10 points.

Michael's Turn

My cards are 4, 2, 7, and 9.
I can't make a 24-hour time.
The latest 12-hour time I can make is 9:47 p.m.

| 9 | : | 4 | 7 | p.m. | | 2 |

Mid-Chapter Review

Frequently Asked Questions

Q: **How can I read the time on a clock?**

A: Look at the hour hand first. On this clock, it's between the 8 and the 9, so the time is after 8:00. Then look at the minute hand. It has moved from the 12 to just past the 7. To see how many minutes have passed since 8 o'clock, count 5, 10, 15, 20, 25, 30, 35, ... 36, 37. The clock shows *eight thirty-seven*, or 8:37. This is also *thirty-seven minutes after eight*, or *twenty-three minutes to nine*.

Q: **How can I read a 24-hour time?**

On the 24-hour clock, 17:30 means 5:30 in the afternoon.

A: The hours on a 24-hour clock and a 12-hour clock are the same from after midnight to 12:00 noon. The hours after 12:00 noon on a 24-hour clock are 13:00, 14:00, 15:00, and so on, and on a 12-hour clock they are 1:00 p.m, 2:00 p.m., 3:00 p.m., and so on.

To read 17:30 as a 12-hour time, just subtract the 12 morning hours. The answer will tell you how many hours have passed since noon.

17 − 12 = 5, so the time is 5:30 p.m.

Q: **How can I read a numeric date?**

A: Numeric dates use numbers to represent the year, the month (01 to 12), and the day (01 to 31). There are different ways to write numeric dates. For example, 2008-06-12 and 12/06/2008 are both ways to write June 12, 2008.

Practice

Lesson 1

1. What will you likely be doing tomorrow at 2:00 p.m.?

Lessons 2, 3, and 4

2. Write each time using numbers and using words.

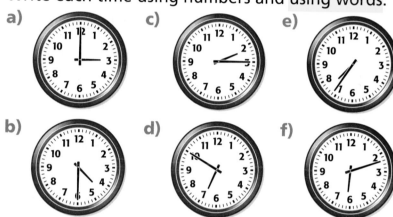

a)   c)   e)

b)   d)   f)

Lesson 5

3. Tom Longboat was a long-distance runner.
   a) He was born on 1887-06-04. The middle
      number shows the month.
      Write the date in calendar form.
   b) He won the Boston Marathon on April 19, 1907.
      Write the date in numeric form.

4. Write each time using 24-hour form.
   a) 8:25 p.m.   b) 11:48 a.m.

5. Write each time using a.m. or p.m.

a)   b)

Chapter 8
Lesson 6

Measuring with Area Units

You will need

- square pattern blocks
- other area units including pattern blocks, pennies, and round counters
- objects to measure (for example, different-sized postcards, bookmarks)

GOAL

Explain why the square is the most efficient unit for measuring area.

Alec noticed that most of the postcards at the store were the same size.

 What can you use to measure the area of a postcard?

Alec's Postcard

I'll measure the **area** of the postcard by covering it with units. Since I'm measuring, I need to use units that are all the same.

A. Measure the area of the postcard with square pattern blocks, and then record the area.

B. Measure the area 2 more times using different units. Record your measurements.

Reflecting

C. Why do you think Alec said he needs to use units that are all the same?

D. Which area unit was the easiest to count? Why?

E. When you measure area, why is it a problem if the units overlap or have gaps between them?

Checking

1. a) Measure the area of this bookmark, and record the measurement.
   b) Which area unit did you use? Why?

Practising

2. a) Choose 2 items whose areas you can measure. Use square pattern blocks to measure the area of one item.
   b) Use your measurement from part (a) to estimate the area of the second item. Record your estimate, and then measure.
   c) Compare the areas you measured. Which area is greater? About how much greater is it?

3. a) Choose 2 different area units. Use one of them to measure the area of the cover of your math book.
   b) Use your measurement from part (a) to estimate what the area would be if you used the other unit. Record your estimate, and then measure.

4. If you wanted to measure the area of a circle, would it be a good idea to use a circle-shaped unit? Why or why not?

5. When you record a measurement, why is it important to include the unit as well as the number?

6. Why might a square be the best shape for measuring area?

Pattern-Block Areas

You will need
• pattern blocks

A square usually makes the best area unit, but not when you're working with pattern blocks.

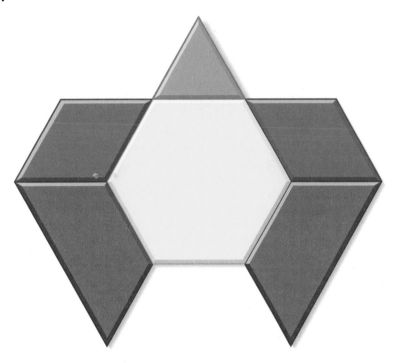

1. How many green triangle blocks do you need to measure the area of this block design? How do you know?

2. a) Use 6 pattern blocks to make a different design. Don't use orange or tan.
   b) What is the area of the design you made?

3. What is the area of the largest possible design you could make with 6 pattern blocks? Explain how you know.

4. Why is the green triangle a good unit to use to measure pattern-block areas?

Counting Square Units

You will need
- grid paper
- pencil crayons

GOAL

Compare and order areas by counting square units.

Aneela and her friends are making designs on square grids. Here are the rules:
- You can use only 2 colours.
- Squares that have the same colour can't touch sides.

Here are 3 of their designs.

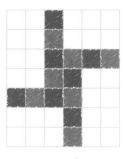
Aneela

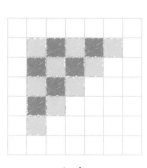
Luis

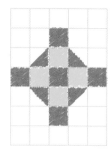
Ethan

 Whose design covers the greatest area?

Aneela's Solution

I can count the **square units** to measure the area of each design and compare the results.

square unit

A square-shaped unit for measuring area

A. Estimate which design has the greatest area.

B. Count the squares covered by the first 2 designs.

C. Count the number of full squares covered by Ethan's design. Some parts of the design are not whole squares. How can you include them when you measure the area?

D. Order the designs from least area to greatest area.

Reflecting

E. How are the shapes used in Ethan's design different from the ones in the other designs?

F. Describe how you figured out the area of Ethan's design.

Checking

1. Renée made this design.
   a) Estimate which covers the most area: yellow, green, or white.
   b) What is the area of each colour?
   c) What is the area of the white part?
   d) Order the areas from least to greatest.

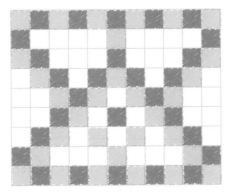

Renée's Design

Practising

2. Jay's design uses letters.

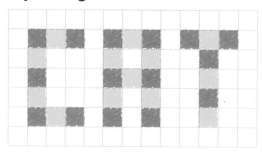

Jay's Design

a) What is the area of each letter?
b) Order the letters from least area to greatest area.
c) What area of the grid is red? yellow? white?
d) Order the 3 colours from least area to greatest area.

3. Allison's design has an area of 16 square units.

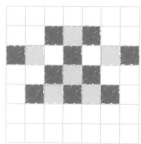

Allison's Design

Colour 2 more designs, each with an area of 16 square units. Use 2 colours in each design, and make sure same-colour squares don't touch sides.

4. Use grid paper to make your own 2-colour design. Use 2 colours and make sure same-colour squares don't touch sides. What is the area of each colour?

5. Colour the initial of your first name and the initial of your last name on grid paper. Which initial has the greater area? Explain how you know.

Area Logic

Number of players: 2 or 3
How to play: Colour areas on a grid. Compare the areas of the colours.

You will need
- 2 dice
- grid paper
- pencil crayons

- **Step 1** The first player rolls 2 dice. The product of the 2 numbers is the area of a rectangle.

- **Step 2** The player colours the rectangle on grid paper.

- **Step 3** The next player rolls 2 dice and colours another rectangle on the same grid. If there is no space left to colour the rectangle, the player loses the turn.

- **Step 4** Continue taking turns until no player can go. Count the coloured squares to see which player has coloured the greatest area.

Olivia's Turn

I rolled 2 and 4. That's an area of 8 squares. I can colour a 2 by 4 or a 1 by 8 rectangle.

$2 \times 4 = 8$

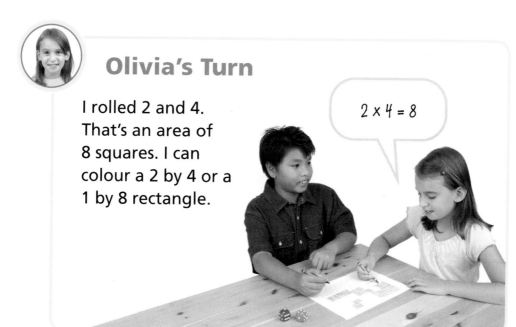

Using Square Centimetres

You will need
- 1 cm grid paper

GOAL

Estimate, measure, and compare areas using square centimetres.

Jade and Ethan are on the lacrosse team.
Each player's picture will be in the yearbook.
They can choose one of these picture sizes.

Jade's Choice

I think the square picture has the greater area.

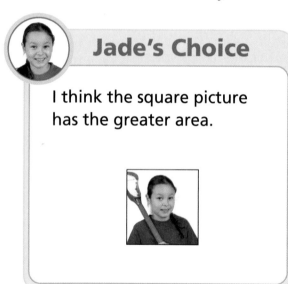

Ethan's Choice

I think the rectangular picture has the greater area.

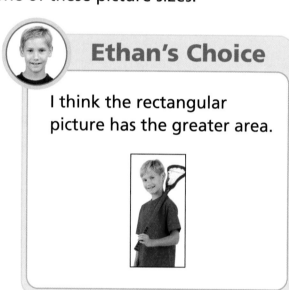

square centimetre (cm²)

A unit of measurement for **area**

1 cm

1 cm

❓ Which picture has the greater area?

A. Which person is right? Explain your reasoning.

B. A **square centimetre (cm²)** is the amount of space covered by a square with sides 1 cm long. It is about the area of your fingernail. About how many square centimetres will fit over each picture? Record your estimates.

C. Place a centimetre grid over each picture. Measure and record the number of square centimetres that each picture covers.

D. Which area is greater? How much greater is it?

Reflecting

E. Why is it hard to compare the picture areas by just looking?

F. Which picture has an area of about 5 cm²? How can the word *about* help you to describe some areas?

G. How is a centimetre grid like a ruler? How is it different?

Checking

1. a) Which picture do you think has the greater area?

A

B

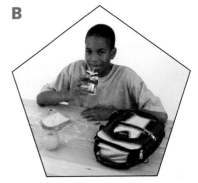

b) Estimate the area of each picture.
c) Measure the area of each picture using a centimetre grid. If there are part squares, use *about* in your answer.

Practising

2. Estimate and then measure each area. Use a centimetre grid and write your answers using cm². If there are part squares, use *about* in your answer.

a)

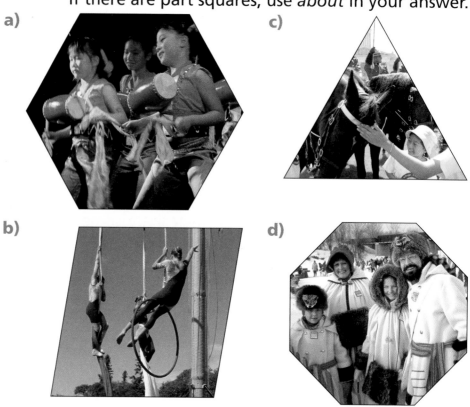

c)

b)

d)

3. a) Which picture do you think has the greater area?

A B

b) Measure the area of each picture. Use a centimetre grid.

4. a) What is the area of the square face of a hundreds block?
   b) How can you use the hundreds block to estimate the area of a piece of paper?
   c) What else could you use to estimate areas in square centimetres?

5. a) Choose 2 surfaces you can measure with a centimetre grid. Estimate each area in square centimetres. Use your estimating ideas from Question 4.
   b) Measure and record each area.

6. When 4 identical squares are joined together so that they share some sides, the shape is called a *tetromino*. There are 5 possible tetrominoes. Here are 2 of them.

 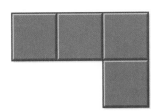

   a) Colour the 5 different tetrominoes on a centimetre grid.
   b) Cut out your tetrominoes. Fit all 5 together to make a shape. What is the area of the shape?
   c) Make another shape with the 5 tetrominoes. What is the area?

7. Luis's school photo is 6 cm wide and 8 cm high. Outline a photo this size on centimetre grid paper. Colour a frame to go around the photo. What is the area of the frame you coloured?

8. Draw a shape that you think covers about 25 cm². Use a centimetre grid to measure your shape.

9. Is a shape with an area of 1 cm² always a square? Use an example to help you explain.

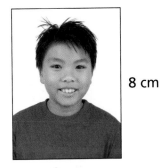

8 cm

6 cm

Lesson 9

Using Square Metres

You will need
- newspaper
- a metre stick
- scissors
- tape

GOAL

Estimate, measure, and compare areas using an appropriate unit.

Carpet is often measured in square metres.

? **About how many square metres of carpet would you need to cover the classroom floor?**

square metre (m²)

A unit of measurement for area

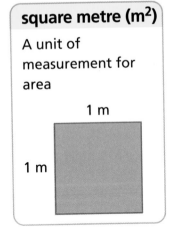

A. Use newspaper to make a model of a **square metre (m²)**. One square metre is the amount of space covered by a square with sides 1 m long.

B. Use your model to estimate the area of the classroom floor.

C. Measure the floor area with your model. About how many square metres of carpet would you need to cover it?

Reflecting

D. How did you measure in Part C? Explain why your measurement could be called an estimate.

E. Suppose you used square centimetres to measure instead of square metres. Would you need more square centimetre units or fewer? Explain using pictures, numbers, and words.

Checking

1. Find a surface that you estimate covers more than 2 m². Measure the area to the nearest square metre.

Practising

2. Which unit would you use to measure each area: square metres or square centimetres? Explain each answer.
   a) a baseball field   **c)** a backyard
   b) a book cover   **d)** an envelope

3. Which would you use to estimate an area in square metres: a floor tile or a beach towel? Describe how you would use it.

4. Name 2 things at home that might have an area of about 1 m². Explain how you estimated.

5. Tam cut his square metre and put the pieces together to make a new shape. What is the area of the new shape? Explain how you know.

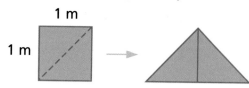

Chapter 8

Lesson 10 Estimating Areas

You will need
- mural paper or sheets of newspaper taped together
- a flashlight

Guess Whose Shadow?
Stephen R. Swinburne

GOAL

Use a square metre for estimating the area of an irregular shape.

Stephen R. Swinburne is a photographer who likes shadows so much that he created this book of shadow pictures.

The size of a shadow depends on the size of the object and the position of the light shining on the object.

 How can you use a flashlight and an object to make a shadow that covers about 1 m²?

296

NEL

Area on Board

A geoboard is like a square grid, except that you can't see the sides of the squares. The area of this geoboard is about 16 square units.
You can divide a 16-square geoboard in half with 3 straight lines.

You will need
- a geoboard
- elastics

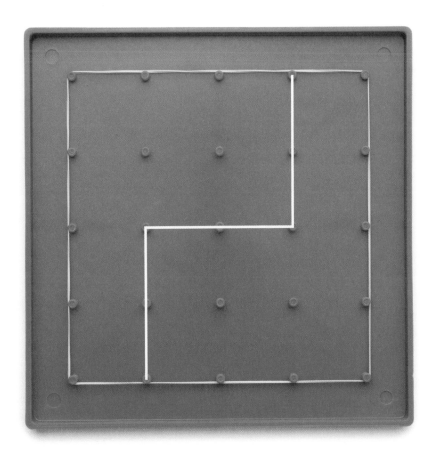

1. How do you know the 2 parts have the same area?

2. What are some other ways to divide a geoboard in half? Sketch pictures to show your ways.

Solving Problems Using Organized Lists

You will need
- square tiles
- 1 cm grid paper

GOAL

Use an organized list to solve problems.

 How many different rectangles with an area of 12 square tiles can you make?

Cory's Solution

Understand

I can make a rectangle that is 1 row of 12 tiles.

I can also make a rectangle that is 2 rows of 6 tiles.

organized list

A way of putting information in order to find all possibilities

Make a Plan

If I keep trying different numbers of rows in order, I should be able to figure out all the rectangles that are possible. I'll keep track using an **organized list**.

Carry Out the Plan

I started with 1 row and then tried all the numbers of rows to 12. I crossed out the rectangles that were the same as another rectangle.

I can make 3 different rectangles with 12 square tiles.

Organized List of Rectangles

1 row of 12 ✔
2 rows of 6 ✔
3 rows of 4 ✔
4 rows of 3 (same as 3 rows of 4)
5 rows (rectangle impossible)
6 rows of 2 (same as 2 rows of 6)
7, 8, 9, 10, 11 rows (rectangle impossible)
12 rows of 1 (same as 1 row of 12)

Reflecting

A. How did Cory's organized list help him find all the different rectangles?

B. How did Cory put his list in order?

Checking

1. How many different rectangles with an area of 36 cm² can you make? Show the steps you use to solve the problem.

Practising

2. How many different rectangles with an area of 40 cm² can you make? Show the steps you use to solve the problem.

3. Carmen has 2 coins in her hand. Together, they are worth less than $1. How much money might she have? Show all the possible answers.

4. Write a problem that can be solved with an organized list. Give your problem to a classmate to solve.

Estimating Areas on Grids

You will need
- 1 cm grid paper
- hundreds blocks

GOAL

Estimate areas of irregular shapes on grids.

The area of skin on a person's body is about 100 times the area of skin on the front of one hand.

 **How many hundreds blocks would it take to show how much skin you have?**

Emily's Solution

I can trace around my hand on centimetre grid paper and count the squares. For every square centimetre on my hand tracing, there are 100 cm² of skin on my body. I can use hundreds blocks to show how much skin I have.

A. Trace your hand on centimetre grid paper.

B. Your hand probably covers some whole squares and some part squares. How many whole squares does it cover?

C. Some squares are half covered or more. How many of those are there?

D. Some squares are less than half covered. How many of those are there?

E. Use your numbers from Parts B, C, and D to estimate the number of hundreds blocks you need to show how much skin you have.

Reflecting

F. Did you trace your fingers close together or spread apart? Would the area change if you did it the other way? Explain.

G. Explain how you estimated. Will your estimate be more or less than the actual area? How do you know?

H. Why is the result called an estimate of the area of your hand rather than a measurement?

Checking

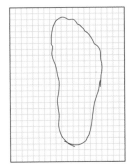

1. Trace your foot on centimetre grid paper. Estimate the area of your foot tracing in square centimetres.

Practising

2. Choose an object and trace around it on centimetre grid paper. Count squares to estimate the area of the tracing.

3. Use your object from Question 2 to estimate a different area in square centimetres. Describe what you did.

4. If an object covers 100 cm^2, is that the same as 1 m^2? Explain how you know.

Chapter Review

Frequently Asked Questions

Q: What is area, and how do you measure it?

A: Area describes the size of a surface. To measure area, you cover the surface with identical area units, usually squares, and count the units. The units are placed next to each other so they don't overlap or leave any gaps.

Q: How can you measure the area of a surface in square centimetres?

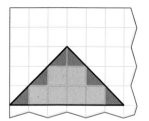

A: You can put a centimetre grid over the area you want to measure. Count the squares the shape covers.

For example, this shape covers 6 whole squares. It also covers 6 half squares.

2 half squares combine to make 1 whole square. The area is 6 + 1 + 1 + 1, or 9 cm².

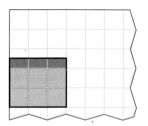

Sometimes you need to just estimate an area. This shape covers 6 whole squares and parts of other squares. The parts combine to make more than 1 square but less than 2 squares. You can estimate that the area is between 7 cm² and 8 cm².

Q: How big is a square metre?

A: A square metre covers the same area as a square that's 1 m long and 1 m wide. For example, that's about the size of 4 desks pushed together.

Practice

Lesson 1

1. Write these times using numbers and a.m. or p.m.
   a) 6 o'clock in the morning
   b) 10 o'clock at night

Lessons 2, 3, and 4

2. Write each time using numbers and using words.

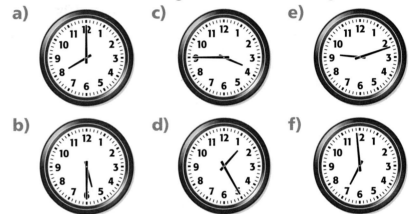

a)   c)   e)

b)   d)   f)

Lesson 5

3. On January 26, 1961, hockey player Wayne Gretzky was born at 7:45 a.m. Write the numeric date and the 24-hour time.

4. Each year on Remembrance Day, November 11th, Canadians pause for 2 minutes of silence at 11:00 a.m. This moment of silence is to remember Canadians who have served in Canada's armed forces. Write the numeric date and 24-hour time that tell when this will happen next.

Lesson 6

5. Which units work better for measuring area: square tiles or pennies? Why?

Reading Strategy

Write statements about what you have learned in this chapter. With a partner, evaluate these statements using *true*, *false*, or *sometimes true*.

Lesson 7

6. Each square has an area of 1 square unit.
   a) Which letter has the least area?
   b) What is the area of the white part of the grid?

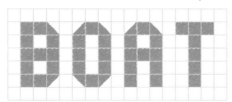

Lessons 8 and 9

7. a) Find 3 surfaces that each cover more than
      10 cm² and less than 1 m².
   b) Use a centimetre grid to measure each area.
   c) Order the 3 areas from least to greatest.

8. Which unit would you use to measure each area:
   square centimetres or square metres? Explain.
   a) a photograph   b) the surface of an ice rink

Lesson 11

9. A rectangle with whole-centimetre sides has an
   area of 32 cm².
   a) Draw all the possible rectangles with this area
      and whole centimetre sides. Label the length
      and width of each rectangle.
   b) Did you find all the rectangles? Explain how
      you know.

Lesson 12

10. Cut a shape out of paper. Estimate the area by
    tracing it on centimetre grid paper.

What Do You Think Now?

Look back at **What Do You Think?** on page 261. How have your
answers and explanations changed?

Chapter Task

✔ Did you include explanations for Parts B, C, and D?

✔ Are your explanations clear?

✔ Did you use math language?

✔ Did you check to make sure the time and date were written correctly?

Making a Photo Display

Olivia cut some pictures she liked out of a magazine. Then she arranged the pictures on centimetre grid paper.

How can you make a photo display?

A. Cut out some pictures. Arrange them on centimetre grid paper.

B. Estimate the area of your largest picture. Explain how you estimated.

C. Estimate the area of your smallest picture. Explain how you estimated.

D. Is more of your grid area covered or not covered? Show how you solved the problem.

E. Write the numeric date and 24-hour time to show when you completed your display.

Multiplying Multi-Digit Numbers

GOALS

You will be able to

- recognize and represent multiplication situations
- multiply 2-digit and 3-digit numbers by 1-digit numbers in different ways
- estimate products
- communicate about solving multiplication problems
- create and solve multiplication problems

How can you use multiplication to count some of the eggs? Use pictures, words, or symbols. Name some other situations where you could count using multiplication.

Getting Started

You will need
- dice
- square tiles
- a multiplication table

Investigating Multiplication Facts

One **array** can represent 2 multiplication facts.

Suppose you roll 2 dice to decide the number of rows and columns for the array.

 How many multiplication facts can you represent?

Ken's Array

I rolled 3 and 6.

I can make an array with 3 rows of 6 tiles.

$3 \times 6 = 18$

$6 \times 3 = 18$

×	0	1	2	3	4	5	6
0	0	0	0	0	0	0	0
1	0	1	2	3	4	5	6
2	0	2	4	6	8	10	12
3	0	3	6	9	12	15	(18)
4	0	4	8	12	16	20	24
5	0	5	10	15	20	25	30
6	0	6	12	(18)	24	30	36

I can also make a 2 by 9 array with 18 tiles.

A. Roll 2 dice.

B. Make an array with tiles. Use the numbers you rolled as the numbers of rows and columns.

C. Write 2 multiplication facts for the array. Circle the **products** on a multiplication table.

D. If possible, rearrange the tiles to make other arrays.

E. Write 2 multiplication facts for each new array. Circle the products on a multiplication table.

F. Repeat Parts A to E 6 times. How many different products can you represent?

G. A multiplication table usually has 10 rows and 10 columns. How do you know there are 100 products in the table?

H. If you kept rolling, how many of the 100 products would you be able to circle? How do you know?

What Do You Think?

Do you *agree* or *disagree* with each statement? Explain your thinking.

1. It's easy to calculate 3×8 if you know that $3 \times 4 = 12$.

2. When you multiply 2 numbers less than 10, the product is usually less than 20.

3. The **product** of two numbers is usually greater than the **sum** of the two numbers.

Exploring Multiplication

GOAL

Use your own strategies to solve everyday math problems.

Your class is having a pizza party.
The pizza comes from Zippy Pizza.

ZIPPY PIZZA
8-Slice Pizza
$9

? **How much will the pizzas cost for your class party?**

MATH GAME

Twenty-Four

You will need
- a spinner with numbers 1 to 9
- a paper clip

Number of players: 2 to 4
How to play: Multiply numbers and compare the product to 24.

- **Step 1** The first player spins the spinner twice and multiplies the 2 numbers.

- **Step 2** Each player takes a turn spinning and multiplying. The player with a product closest to 24 gets a point.

- **Step 3** Players continue taking turns and multiplying and scoring points at the end of each round.

The first player with 10 points wins.

 ## Hailey's and Annie's Spins

I spun 4 and 8.

I spun 5 and 4.

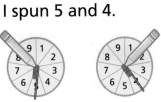

$4 \times 8 = 32$

$5 \times 4 = 20$

20 is closer to 24 than 32 is.
Annie wins a point.

Chapter 9
Lesson 2

Multiplying 10s and 100s

You will need
- base ten blocks

GOAL

Use patterns to multiply 10s and 100s.

Diane is making safety pin bracelets.
She uses 100 beads and 50 safety pins for each.

? How many pins and beads does Diane need for 5 bracelets?

Diane's Calculations

5 × 100 is the number of beads.
5 × 50 is the number of pins.
First, I'll model 5 × 100.
5 × 100 is 5 groups of 100.

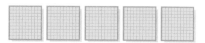

A. Complete a table like this for up to 5 bracelets.

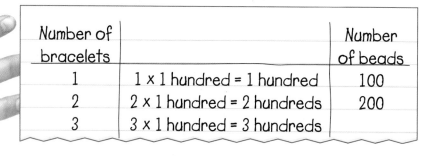

Number of bracelets		Number of beads
1	1 × 1 hundred = 1 hundred	100
2	2 × 1 hundred = 2 hundreds	200
3	3 × 1 hundred = 3 hundreds	

B. What do you notice about the number of beads each time?

312

NEL

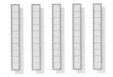

C. Model the number of pins.
Complete a table like this for up to 5 bracelets.

Number of bracelets		Number of pins
1	1 x 5 tens = 5 tens	

D. How many beads and pins does Diane need?

Reflecting

> **E.** What patterns do you see in your tables?
>
> **F.** Why does it make sense that there will be a 0 in the ones place when you multiply 10s or 100s?

Checking

1. How many beads and pins does Diane need for 7 bracelets?

Practising

2. Multiply.
   a) 7×10 b) 3×40 c) 50×4 d) 9×200

3. Darren's family took 3 trips each year for 10 years. How many trips did his family take in total?

4. What is the missing number?
   a) $400 = \blacksquare \times 100$ c) $80 = \blacksquare \times 10$
   b) $60 = \blacksquare \times 10$ d) $700 = 7 \times \blacksquare$

5. Charlie's family bought an old sled with six $20 bills. How much did the sled cost?

6. You are telling a friend how to multiply 6 by 30 and by 300. What do you say?

Lesson 3 Multiplying Using Arrays

You will need
- grid paper
- pencil crayons

 How can you calculate the number of spaces on Alec's 8-by-12 game board?

Alec's Solution

My game board has 8 rows of 12 spaces.
I'll sketch it on grid paper.
8 × 12 tells the number of spaces.
I know 8 × 10 already, so I'll split the 8-by-12 array into an 8-by-10 array and an 8-by-2 array.

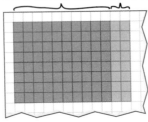

8 rows of 10   8 rows of 2

$8 \times 12 = 8 \times 10 + 8 \times 2$

A. Use $8 \times 10 + 8 \times 2$ to calculate 8×12.

B. Sketch another 8-by-12 array on grid paper.
This time, colour an 8-by-8 array.
Then colour the leftover 8-by-4 array.

C. Write 8×12 as the sum of 2 products
to represent what you coloured.
$8 \times 12 = 8 \times \blacksquare + 8 \times \blacksquare$
$8 \times 12 = \blacksquare + \blacksquare$
$8 \times 12 = \blacksquare$

D. Sketch another 8-by-12 array on grid paper.
Visualize 2 different arrays inside it. Colour them.
Show your work as you did in Part C.

E. How many spaces are on Alec's game board?

Reflecting

F. How does splitting an array into smaller arrays show you that there is more than one way to calculate a product?

G. What other ways can you visualize an 8-by-12 array to calculate 8×12?

Checking

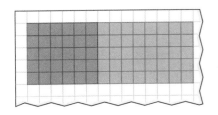

1. a) Complete the number sentence to show how the array is coloured.
$5 \times 14 = 5 \times \blacksquare + 5 \times \blacksquare$
   b) Calculate 5×14.

2. A game board has 7 rows of 14 squares.
   a) Sketch the array on grid paper. Colour smaller arrays that have easier numbers to multiply.
   b) Write 7×14 as the sum of 2 products. How many squares are on the game board?

Practising

3. Complete.

a)

$6 \times 12 = 6 \times 10 + 6 \times \blacksquare$

$6 \times 12 = \blacksquare + \blacksquare$

$6 \times 12 = \blacksquare$

b)
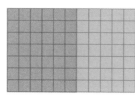

$7 \times 11 = 7 \times \blacksquare + \blacksquare \times \blacksquare$

$7 \times 11 = \blacksquare + \blacksquare$

$7 \times 11 = \blacksquare$

c)

$5 \times 27 = \blacksquare \times \blacksquare + \blacksquare \times \blacksquare$

$5 \times 27 = \blacksquare + \blacksquare$

$5 \times 27 = \blacksquare$

4. Jiri planted 7 rows of 18 trees.
   How many trees did he plant?

5. Complete.

a) $6 \times 21 = 6 \times 20 + 6 \times 1$

   $6 \times 21 = \blacksquare$

b) $4 \times 16 = 4 \times 8 + 4 \times \blacksquare$

   $4 \times 16 = \blacksquare$

c) $5 \times 32 = 5 \times \blacksquare + 5 \times \blacksquare$

   $5 \times 32 = \blacksquare$

d) $5 \times 28 = \blacksquare \times \blacksquare + \blacksquare \times \blacksquare$

   $5 \times 28 = \blacksquare$

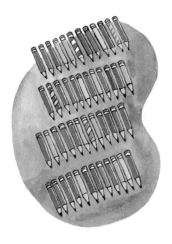

6. Calculate the total in each array.
   a) 5 rows of 21 carrots
   b) 6 rows of 18 trading cards
   c) 4 rows of 12 pencil crayons
   d) 9 rows of 17 new cars
   e) 8 rows of 16 bottles
   f) 3 rows of 19 stamps

7. Multiply.
   a) $3 \times 17 =$ ▢
   b) $2 \times 16 =$ ▢
   c) $4 \times 15 =$ ▢
   d) ▢ $= 8 \times 14$
   e) ▢ $= 9 \times 14$
   f) ▢ $= 9 \times 23$

Reading Strategy

What do these rows in the greenhouse look like in your mind?

8. A greenhouse has
   3 rows of 11 vegetables,
   3 rows of 13 vegetables,
   3 rows of 18 flowers, and
   3 rows of 16 flowers.
   Are there more vegetables or more flowers?

9. Sketch arrays to show that each statement is true.
   a) $5 \times 23 = 5 \times 20 + 5 \times 3$
   b) $5 \times 23 = 5 \times 10 + 5 \times 10 + 5 \times 3$
   c) $5 \times 23 = 5 \times 7 + 5 \times 7 + 5 \times 7 + 5 \times 2$

10. Explain how you might calculate 4×19 as
    $4 \times 20 - 4 \times 1$. Use an array.

11. Which methods can you use to calculate 7×15?
    Explain.
    A. $7 \times 10 + 7 \times 5$
    B. $7 \times 5 + 7 \times 5 + 7 \times 5$
    C. $7 \times 5 \times 7 \times 3$
    D. $7 \times 7 + 7 \times 7 + 7 \times 1$

Multiplying Using Expanded Form

You will need
- base ten blocks
- a place value chart

GOAL

Multiply 2-digit numbers by 1-digit numbers using expanded form.

Annie is making 54 leather bags for gifts at a celebration.
She sews 3 designs on each bag.

? **How many designs will Annie sew?**

Annie's Solution

54 bags with 3 designs each is 54 groups of 3.
$54 \times 3 = 3 \times 54$
I'll use 3×54 to show my work.

- **Step 1** I'll make 3 groups of 54 with base ten blocks.
  I'll record the numbers in expanded form.

$$3 \times 54 = \blacksquare$$

Hundreds	Tens	Ones

$$\begin{array}{r} 54 \\ \times\ 3 \\ \hline \end{array} \qquad \begin{array}{r} 50 + 4 \\ \times\ 3 \\ \hline \end{array} \qquad \begin{array}{r} 5 \text{ tens} + 4 \text{ ones} \\ \times\ 3 \\ \hline \end{array}$$

- **Step 2**   I see 3 groups of 5 tens, or 15 tens.

$$
\begin{array}{r}
54 \\
\times\,3 \\
\hline
150
\end{array}
\qquad
\begin{array}{r}
50 + 4 \\
\times\,3 \\
\hline
150
\end{array}
\qquad
\begin{array}{r}
5 \text{ tens} + 4 \text{ ones} \\
\times\,3 \\
\hline
15 \text{ tens}
\end{array}
$$

- **Step 3**   I see 3 groups of 4 ones, or 12 ones.

$$
\begin{array}{r}
54 \\
\times\,3 \\
\hline
150 \\
+\,12 \\
\hline
\end{array}
\qquad
\begin{array}{r}
50 + 4 \\
\times\,3 \\
\hline
150 \\
+\,12 \\
\hline
\end{array}
\qquad
\begin{array}{r}
5 \text{ tens} + 4 \text{ ones} \\
\times\,3 \\
\hline
15 \text{ tens} \\
+\,12 \text{ ones} \\
\hline
\end{array}
$$

- **Step 4**   I'll regroup the blocks.
  I'll add the 2 products.

Hundreds	Tens	Ones

$$
\begin{array}{r}
54 \\
\times\,3 \\
\hline
150 \\
+\,12 \\
\hline
162
\end{array}
\qquad
\begin{array}{r}
50 + 4 \\
\times\,3 \\
\hline
150 \\
+\,12 \\
\hline
162
\end{array}
\qquad
\begin{array}{r}
5 \text{ tens} + 4 \text{ ones} \\
\times\,3 \\
\hline
15 \text{ tens} \\
+\,12 \text{ ones} \\
\hline
162
\end{array}
$$

$$3 \times 54 = 162$$

I'll sew 162 designs.

Reflecting

A. Why do you think Annie decided to show 3 groups of 54 instead of 54 groups of 3?

B. Why was it helpful to calculate 3×50 and 3×4 separately?

Checking

1. How can you model and calculate 5×67?

2. Sam serves 4 trays of salmon at a potlatch. Each tray holds 32 pieces of salmon. How many pieces of salmon does Sam serve?

Practising

3. Copy and complete.

   a)
   $$\begin{array}{r} 6 \text{ tens} + 2 \text{ ones} \\ \times\ 8 \\ \hline 48 \text{ tens} \\ +\ \blacksquare\ \text{ones} \\ \hline \blacksquare\blacksquare\blacksquare \end{array}$$

   $8 \times 62 = \blacksquare$

   b)
   $$\begin{array}{r} 70 + 4 \\ \times\ 6 \\ \hline 420 \\ +\ \blacksquare\blacksquare \\ \hline \blacksquare\blacksquare\blacksquare \end{array}$$

   $6 \times 74 = \blacksquare$

4. A box holds 5 pencils. How many pencils are in 85 boxes?

5. Alasie made a bracelet with 6 rows of 64 beads.
   a) How do you think Alasie knew she would need more than 350 beads?
   b) How many beads did she use altogether?

6. Multiply.
   a) 5×17 b) 7×19 c) 6×56 d) 8×25

7. Each shelf holds 21 books.
   How many books are on 4 shelves?

8. Multiply.
   a) 4×73 b) 2×72 c) 3×68 d) 2×92

9. Ava is playing Scrabble. She adds the letter values to get a word score.
   She can get a triple word score (3 times the regular score) with

   or a double word score (2 times the regular score) with

   Which word scores more points?

10. Multiply.
    a)   29   b)   16   c)   28   d)   87
       $\times\,5$ $\times\,3$ $\times\,6$ $\times\,3$

11. $4 \times \blacksquare$ = 16 tens + 12 ones
    What is the value of $\blacksquare$? How do you know?

12. I used 27 tens and ones blocks to model $3 \times \blacksquare$.
    What are 3 possible values for $\blacksquare$?
    How do you know?

13. When you multiply a 2-digit number by a 1-digit number, how many digits can be in the answer? Explain.

14. How is adding 25 to 8 different from multiplying 25 by 8?

Estimating Products

Estimating Products

GOAL

Choose when and how to estimate.

Eight soccer teams competed in a tournament.
The chart shows the number of players on each team.

Number of Players

Red Deer	17	17	17	17
Calgary	20	19	19	18

Michael was filling out an information sheet for the home coach.

Soccer Tournament Information

1. Are there more than 80 players from Red Deer? _____
2. How many players are there in total? _____

 When can Michael estimate and when does he have to calculate?

Michael's Solution

I can estimate to answer Question 1,
but I have to calculate to answer Question 2.
To estimate 17 + 17 + 17 + 17, I can multiply
20 by 4.

A. Why can Michael estimate to answer Question 1?

B. Answer Question 1. Explain your thinking.

C. Why can't Michael estimate to answer Question 2?

D. Michael answered Question 2 by writing 164. How can you estimate to determine that his answer is wrong?

E. Make up another question for the information sheet that you could answer by estimating.

F. Make up a question for the information sheet that you could answer *only* by calculating.

Reflecting

G. Why does Michael's estimate make sense?

H. How did you decide whether you could estimate, or not, to answer a question?

Checking

1. Natasha's school has 2 Grade 5 teams. Each team has 31 players. Can you estimate to answer the question below? Explain.

> Are there more than 50 Grade 5 players?

2. How would you estimate each product?
   a) 9×48 b) 4×355

Practising

3. How would you estimate each product?
   a) 6×132 b) 7×93

4. Which products are about 500?
   How do you know?
   A. 6 × 82   **B.** 4 × 92   **C.** 4 × 216   **D.** 9 × 53

5. 48 mushers entered a dog-sled race.
   Each musher had 6 dogs.
   About how many dogs were in the race?

6. Decide whether you can estimate to answer.
   Then answer.
   a) 72 minutes of music can be burned onto one
      CD. Are 7 CDs enough to burn 500 minutes of
      music?

   b) Each plate contains 76 dumplings. There are
      3 plates. Are there at least 200 dumplings?

   c) Jonah has $287 in his bank account.
      His brother saved 3 times as much money.
      Did his brother save at least $900?

7. Estimate to answer each question. Explain how
   you estimated.
   a) Is 6 × 53 equal to 308?
   b) Is 4 × 87 equal to 388?
   c) Is 4 × 165 equal to 480?
   d) Is 5 × 88 equal to 400?

8. Why might someone estimate to predict an
   answer?

9. Why might someone estimate to check an answer?

Greatest Product

You will need
- a spinner with numbers 1 to 9
- a paper clip

Number of players: 2 to 4
How to play: Create the greatest product.

- **Step 1** Each player records boxes like this.

  ☐ × ☐☐

- **Step 2** The first player spins the spinner 3 times. After each spin, the player records the digit in one of the boxes. Then the other players spin and record.

- **Step 3** All players multiply and compare products.

The player with the greatest product wins.

Hailey's Turn

My spins were 4, 3, and 7.
I'll use base ten blocks to multiply.
My product is 148.

4 × 3 7

325

Mid-Chapter Review

Frequently Asked Questions

Q: **How can you multiply a 2-digit number by a 1-digit number?**

A1: You can use arrays. For example, for 4×27 you can use a 4 by 27 array and add the parts of the array.

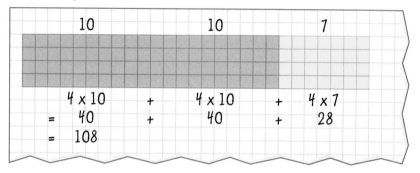

A2: You can use expanded form.
For example, $4 \times 27 = $ ▨

$$
\begin{array}{r}
2 \text{ tens} + 7 \text{ ones} \\
\times\ 4 \\
\hline
8 \text{ tens} \\
+\ 28 \text{ ones} \\
\hline
108
\end{array}
$$

$$4 \times 27 = 108$$

Q: **How can you estimate the product of a 2-digit number and a 1-digit number?**

A: You can use close numbers that are easier to multiply. For example, to estimate 6×38, you can use $6 \times 40 = 240$.

Practice

1. Which problem can be solved using multiplication? Explain how you know.
   - A. A basketball player played in 3 games. In each game, he scored 22 points. How many points did he score altogether?
   - B. A basketball player scored 26 points. Another player scored 32 points. How many points did they score together?

Lesson 2
2. Calculate.
   - a) 6×10
   - b) 5×20
   - c) 3×100
   - d) 7×800

Lesson 3
3. Olivia made a 6 by 16 tray of baklava. How many pieces did she make? Use a sketch and a number sentence in your solution.

   $6 \times 16 = \square \times \square + \square \times \square$

Lesson 4
4. Calculate each product. Show your work.
   - a) 3×18
   - b) 6×11
   - c) 4×71
   - d) 5×17

5. Lingling is making 19 red paper fans for Chinese New Year. Each fan is made with 7 chopsticks. How many chopsticks will Lingling need? Show your work.

6. Ryan is making 27 beanbag creatures. Each creature will have 3 buttons on its face. How many buttons does Ryan need? Show your work.

Lesson 5
7. Justine estimates that 8×72 is more than 560. How do you think she chose her estimate?

Communicating about Solving Problems

 GOAL

Explain your thinking when solving a problem.

Horses age more quickly than humans. For every year a horse lives, it ages about 3 human years.
Ken wondered how old his 16-year-old horse would be in human years.

 How can Ken explain how he solved the problem?

Ken's Solution and Explanation

I asked Desmond to help me improve my explanation.

> You showed the problem-solving steps.

1. I made sure I understood the problem.

A horse this old...	...is like a human this old
1	3
2	6
3	9

> How did you know you could multiply?

> What numbers did you multiply?

2. I made a plan. I'll multiply.
3. I carried out the plan. The answer is 48.
   My horse is like a 48-year-old human.

> You looked back to check your answer.

4. Then I looked back to check.
   I think 48 years old is a reasonable answer.

> How do you know 48 is reasonable?

A. What did Desmond think was good about Ken's explanation?

B. What did Desmond think was missing from Ken's explanation?

C. How would you answer Desmond's questions?

Communication Checklist

✔ Did you show the right amount of detail?
✔ Did you explain your thinking?

Reflecting

D. How do Desmond's questions relate to the questions in the Communication Checklist?

Checking

1. For every year a bear lives, it ages about 4 human years. Carolyn calculated the age of a 19-year-old bear in human years.

1. I made sure I understood the problem.

A bear this old...	...is like a human this old
1	4
2	8
3	12

2. I made a plan. I calculated 19 × 4.
3. I carried out the plan. 19 × 4 = 76.
4. I looked back to check. 76 looks right because 20 × 4 = 80, so 19 × 4 must be less.

a) What did Carolyn explain well?
b) What questions would you ask Carolyn to help her improve her explanation?

Practising

2. For every year a dog lives, it ages about 7 human years. How old is a 13-year-old dog in human years? Solve the problem and explain your solution.

Chapter 9
Lesson 7

Multiplying 3-Digit Numbers

You will need
- base ten blocks
- a place value chart

GOAL

Multiply 3-digit numbers by 1-digit numbers using expanded form.

Diane lives near a beach.
She collected 114 shells one week.
She wants to collect the same number each week.

 How many shells will Diane have in 4 weeks?

Diane's Solution

I'll have 4 groups of 114 after 4 weeks.
When there are equal groups, I can multiply.

- **Step 1** I'll estimate first.
  4×114 is about $4 \times 100 = 400$.
  I predict that I'll have more than 400 shells.

- **Step 2** I'll make 4 groups of 114 with base ten blocks.
  I'll record using expanded form.

Thousands	Hundreds	Tens	Ones
	▨	▯	▫ ▫ ▫ ▫
	▨	▯	▫ ▫ ▫ ▫
	▨	▯	▫ ▫ ▫ ▫
	▨	▯	▫ ▫ ▫ ▫

$$\begin{array}{r} 114 \\ \times\,4 \\ \hline \end{array} \qquad \begin{array}{r} 100 + 10 + 4 \\ \times\,4 \\ \hline \end{array}$$

- **Step 3** I'll multiply to show the number of hundreds first.

$$\begin{array}{r} 114 \\ \times\ 4 \\ \hline \end{array}$$

$$\begin{array}{r} 100 + 10 + 4 \\ \times\ 4 \\ \hline 400 \\ \blacksquare\blacksquare \\ +\ \blacksquare\blacksquare \\ \hline \blacksquare\blacksquare\blacksquare \end{array}$$

A. Complete Diane's multiplication.

B. How many shells will Diane have?

Reflecting

C. Diane showed the hundreds first. Suppose you multiplied by showing the ones first, like this. Would the product be the same? Explain.

$$\begin{array}{r} 100 + 10 + 4 \\ \times\ 4 \\ \hline 16 \\ \blacksquare\blacksquare \\ +\ \blacksquare\blacksquare\blacksquare \\ \hline \blacksquare\blacksquare\blacksquare \end{array}$$

D. Compare your final product from Part A with Diane's estimate of about 400. Was the product greater or less? Explain why.

Checking

1. Model with base ten blocks. Multiply.

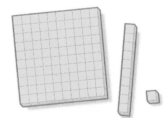

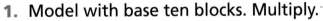

a)
$$\begin{array}{r} 300 + 20 + 7 \\ \times\ 5 \\ \hline 1500 \\ 100 \\ +\ \blacksquare\blacksquare \\ \hline \blacksquare\blacksquare\blacksquare\blacksquare \end{array}$$

b)
$$\begin{array}{r} 327 \\ \times\ 5 \\ \hline 35 \\ \blacksquare\blacksquare\blacksquare \\ +\ \blacksquare\blacksquare\blacksquare\blacksquare \\ \hline \blacksquare\blacksquare\blacksquare\blacksquare \end{array}$$

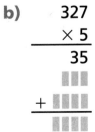

2. The students in René's class read 146 books in September, 81 in French and 65 in English. They plan to read the same number each month. How many books will they read in 6 months?

Practising

3. Multiply.

a)
$$
\begin{array}{r}
200 + 80 + 3 \\
\times\ 7 \\
\hline
1400 \\
\blacksquare\blacksquare\blacksquare \\
+\ \blacksquare\blacksquare \\
\hline
\blacksquare\blacksquare\blacksquare\blacksquare
\end{array}
$$

b)
$$
\begin{array}{r}
400 + 60 + 9 \\
\times\ 9 \\
\hline
\blacksquare\blacksquare \\
\blacksquare\blacksquare\blacksquare \\
+\ \blacksquare\blacksquare\blacksquare\blacksquare \\
\hline
\blacksquare\blacksquare\blacksquare\blacksquare
\end{array}
$$

4. Kristi plants 350 carrots each month from May to July. How many carrots does she plant altogether?

5. Calculate.
a) 7×361 b) 4×421 c) 3×618 d) 6×333

6. 485 students will each carry 3 balloons in a school parade. 125 students will each carry 2 flags. How many balloons and flags will they carry?

7. Estimate, then calculate.
a) 3×986 b) 5×181 c) 7×332 d) 5×885

8. Each kilogram of trail mix contains 340 g of Saskatoon berries. How many grams of Saskatoon berries are needed for 7 kg of trail mix?

9. Multiply.

a)
$$
\begin{array}{r}
125 \\
\times\ 8 \\
\hline
\end{array}
$$

b)
$$
\begin{array}{r}
753 \\
\times\ 5 \\
\hline
\end{array}
$$

c)
$$
\begin{array}{r}
618 \\
\times\ 7 \\
\hline
\end{array}
$$

d)
$$
\begin{array}{r}
275 \\
\times\ 3 \\
\hline
\end{array}
$$

10. Create a problem you can solve by multiplying 8 by 217. Solve your problem.

11. How is multiplying a 3-digit number by 5 like multiplying a 2-digit number by 5? How is it different?

Egyptian Multiplication

Here is a curious way to multiply.
It's called Egyptian multiplication.

Suppose you want to multiply 314.
- Start a list with $1 \times 314 = 314$.
- Double the red number and the product to make each new row in the list.
- Decide what number you want to multiply by 314. Find red numbers that add up to your number. Add the numbers to the right of the equal sign from those rows.

$$1 \times 314 = 314$$
$$2 \times 314 = 628$$
$$4 \times 314 = 1256$$
$$8 \times 314 = 2512$$

$\times 2$

For example, $9 = 8 + 1$, so
$$9 \times 314 = 8 \times 314 + 1 \times 314$$
$$9 \times 314 = 2512 + 314$$
$$9 \times 314 = 2826$$

1. Calculate. Show your work.
   a) 7×314 b) 6×314

CURIOUS MATH

Sum and Product

1. What 3 numbers have a sum equal to their product?
   ▇ + ▇ + ▇ = ▇ × ▇ × ▇

2. 5 different numbers have a sum of 15.
   Those same 5 numbers have a product of 120.
   What are the 5 numbers?

Multiplying Another Way

You will need
- base ten blocks
- a place value chart

> **GOAL**
>
> Multiply, regrouping as you go.

Michael has 256 hockey cards. Pedro has twice as many.

 How many hockey cards does Pedro have?

Michael's Calculations

Twice as many means 2 times as many.
I'll multiply my number of cards by 2.

- **Step 1** I'll estimate first.
  256 is about halfway between 200 and 300.

  $2 \times 200 = 400$ $2 \times 300 = 600$

  500 is halfway between 400 and 600.
  Pedro has about 500 cards.

- **Step 2** I'll multiply by making 2 groups of 256.

Thousands	Hundreds	Tens	Ones

$2 \times 256 = \blacksquare$

```
  2 5 6
x     2
```

- **Step 3** I see 2 × 6 ones. I'll regroup 12 ones as 1 ten, 2 ones.

Thousands	Hundreds	Tens	Ones

```
      1
  2 5 6
x     2
------
      2
```

- **Step 4** I see 2 × 5 tens + 1 ten. That's 11 tens.
  I'll regroup 11 tens as 1 hundred, 1 ten.

Thousands	Hundreds	Tens	Ones

```
  1 1
  2 5 6
x     2
------
    1 2
```

- **Step 5** I see 2 × 2 hundreds + 1 hundred.
  That's 5 hundreds.

  Pedro has 512 hockey cards.

```
  1 1
  2 5 6
x     2
------
  5 1 2
```

Reflecting

A. In Step 3, why did Michael record a number 1 above the number 5?

B. Mark has 756 cards. Shani has twice as many. How do you know the product of 756 and 2 is greater than 1000?
How will the block model and the recording be different from Michael's?

Checking

1. Calculate.
   a) 7×62
   b) 7×145

2. Chantal has 145 plastic horses.
   She has 3 times as many other animals.
   How many other animals does she have?
   Estimate, then calculate the answer.

Practising

3. Estimate. Calculate only those with a product less than 3000.
   a) 3×491 b) 6×952 c) 9×611 d) 9×937

4. Talima and Mark are making dreamcatchers.
   Mark uses 880 cm of sinew.
   Talima uses 3 times as much sinew.
   How much sinew does Talima use?

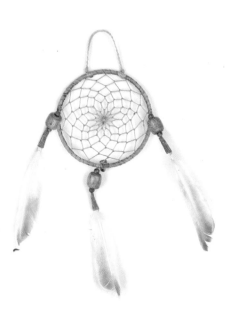

5. Write the multiplication equation for this model.
   Calculate the product.

Thousands	Hundreds	Tens	Ones

6. Jamal calculated 4×384.
   He used 4×400 for his estimate.
   a) Was his estimate high or low? How do you know?
   b) What other 3-digit by 1-digit multiplication calculations could be estimated using 4×400?
   Write 3.

7. Calculate.
   a) 4×305 b) 5×260 c) 6×293 d) 4×429

8. Some jets can carry 131 passengers. A jumbo jet can carry 4 times as many passengers.
   How many passengers can a jumbo jet carry?

9. Multiply.

   a) 225
       $\times 9$

   b) 863
       $\times 3$

   c) 594
       $\times 7$

   d) 943
       $\times 8$

10. Place the digits 2, 4, 6, and 8 in spaces like this.
    What is the greatest product you can make if you use each digit once?

11. Jeff will be 9 years old in 159 weeks.
    a) About how old is Jeff?
    b) About how many days away is his 9th birthday?

12. Make up a 3-digit by 1-digit multiplication problem about a hobby or game that interests you. Solve your problem.

13. Compare Michael's way of multiplying to the one using expanded form. How are they alike? How are they different?

1	1	
2	5	6
x		2
5	1	2

```
  200 + 50 + 6
         × 2
─────────────
          12
         100
       + 400
─────────────
         512
```

Choosing a Method for Multiplying

You will need
• base ten blocks

GOAL

Choose whether to estimate or calculate, and explain your multiplication method.

Did you know that in 2004
• 33 babies were born in Saskatchewan *every 24 hours*, and
• 44 babies were born in Alberta *every 10 hours?*

? How can you solve each problem?

For each situation, decide which method you would use and explain why.
• estimate
• calculate using mental math
• calculate using materials

A. About how many Saskatchewan babies were born in a week?

B. Were more than 500 babies born in Saskatchewan each month?

C. How many Alberta babies were born in 50 hours?

D. About how many Alberta babies were born in a day?

E. How long did it take for 11 babies to be born in Saskatchewan?

Reflecting

F. In what types of situations would you use mental calculations?

G. In what types of situations would you use estimation?

Checking

1. Would you estimate or calculate? Why?
   a) How many Saskatchewan babies were born in 5 days?
   b) Were more than 1000 babies born in Alberta each month?

2. How would you calculate in each situation?
   a) the number of Saskatchewan babies born in 6 days
   b) the number of Alberta babies born in 90 hours

Practising

3. Would you answer each question using mental math or base ten blocks? Explain.
   a) There are 250 sheets in one pack of paper. Are there more than 500 sheets in 3 packs?
   b) How many days are there in 2 years?
   c) Aaron has 3 times as much money as Raven. Does he have more or less than $400?

4. Mary can walk 67 m in a minute.
   Would you estimate or calculate for each? Why?
   a) How far can Mary walk in 8 minutes?
   b) Can she walk 500 m in 5 minutes?
   c) Can she walk 500 m in 8 minutes?

5. Evan earns $7 an hour babysitting. For which of these times would you use mental math to calculate the amount he earns? Why?
   A. 8 hours   **B.** 10 hours   **C.** 125 hours

6. Alana earns $9 an hour babysitting. Which could you answer by estimating?
   A. the amount Alana earns in 10 hours
   B. the amount Alana earns in 15 hours
   C. the number of hours needed to earn $90
   D. about how long it would take Alana to earn $250

7. The chart shows the number of births *each day* in one recent year.

Nunavut	Manitoba	British Columbia
2	38	108

   Would you estimate or calculate to answer each question? Why?
   a) How many Nunavut babies were born in 100 days?
   b) Were more than 1000 babies born in British Columbia each week?
   c) About how many Manitoba babies were born in the winter?

8. Alexis's birthday is 10 to 20 weeks away. She calculated the number of days using mental math. How many weeks away do you think Alexis's birthday is? Why?

340

Creating Multiplication Problems

You will need
• pencil crayons

> **GOAL**
>
> Create and solve multiplication problems.

Alec loved the book *Amanda Bean's Amazing Dream* by Cindy Neuschwander. In that book, Amanda used multiplication to count large numbers of things, from sheep on bicycles to balls of yarn.

Alec wrote his own book.
Each page had a multiplication question on it.
The words on the page went with the multiplication question and included the product.
All together, the pages told a story.
He read the book to his classmates and explained how he calculated.

7 × 45

Kelly practised piano 45 minutes

a day every day of the week.

That makes 315 minutes.

How can you create a story about multiplication?

Chapter Review

Frequently Asked Questions

Q: How can you multiply a 3-digit number by a 1-digit number?

A: You can model with base ten blocks and record. For example, 5 × 113 means 5 groups of 113.

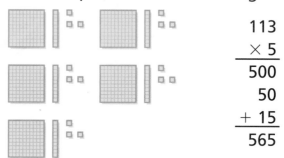

$$\begin{array}{r} 113 \\ \times\ 5 \\ \hline 500 \\ 50 \\ +\ 15 \\ \hline 565 \end{array}$$

	1	
1	1	3
×		5
5	6	5

Q: How do you decide whether to multiply using mental math or using base ten blocks?

A: Which way you multiply might depend on the numbers or what you're comfortable with. For example, for 4 × 100, you would probably use mental math since 4 × 100 is 4 hundreds. For 3 × 254, you might use blocks since there is a lot of regrouping.

Q: Why might you estimate a product instead of calculating it?

A: You might estimate if you want to predict the answer or to check the answer. You might also estimate if you don't need an exact answer. For example, you can estimate to decide if $100 is enough money to buy 3 games that cost $19 each.

Practice

Lesson 2

1. What is the missing number?
   a) $600 = \boxed{} \times 100$
   b) $\boxed{} \times 60 = 480$
   c) $\boxed{} \times 300 = 1800$

Lesson 3

2. Matthew multiplied using this array.

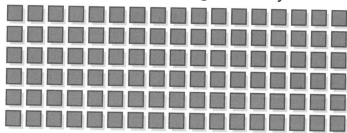

   a) What multiplication does the array show?
   b) Show the array as 2 smaller arrays.
   c) Write the multiplication as the sum of 2 products.

Lesson 4

3. Eric's class has 28 students. Each student donated 5 items for a food drive. How many items did the class donate?

Lesson 5

4. Which estimate is closer to the actual product? Explain your choice.
   a) 79×9 80×10 or 80×9?
   b) 47×7 50×7 or 47×10?

5. Would you estimate or calculate? Why? If you would estimate, explain how.
   a) One rack contains 500 DVDs. How many DVDs are on 3 racks the same size?
   b) Keegan has $187. Her sister has about 3 times as much. About how much money does Keegan's sister have?

6. Jack is 7 years old today. How many days old is he?

7. Calculate.
    a) 3×49 b) 8×77 c) 5×176 d) 6×409

8. Sound travels 344 m in a second.
   How far does sound travel in 7 seconds?

9. Pierre is fencing off a square field for his horse.
   Each side of the field is 152 m long.
   What is the length of the fence?

10. 750 can be represented as 3×250.
    Represent 750 in 3 other ways using products.

11. Multiply. Estimate to check your answers.
    a) 9×35 b) 3×654 c) 4×185 d) 7×348

12. Determine the missing digits.
    Show your calculations.

    a)
    $$\begin{array}{r} \blacksquare 8 \\ \times\ 8 \\ \hline 30\blacksquare \end{array}$$

    b)
    $$\begin{array}{r} 74 \\ \times\ \blacksquare \\ \hline 222 \end{array}$$

    c)
    $$\begin{array}{r} \blacksquare 24 \\ \times\ 5 \\ \hline 26\blacksquare\blacksquare \end{array}$$

    d)
    $$\begin{array}{r} \blacksquare 8\blacksquare \\ \times\ 4 \\ \hline 1128 \end{array}$$

13. Heather says she calculated 2×104 faster using
    mental math than using paper and pencil or a
    calculator. Explain why this could be true.

14. Multiply. Explain your choice of method.
    a) 50×4 b) 312×6 c) 8×479 d) 3×896

What Do You Think Now?

Look back at **What Do You Think?** on page 309. How have your
answers and explanations changed?

Chapter Task

Describing a School Year

Olivia's school has 3 recesses each day, including lunch. There are about 190 days in one school year. So, there are about $3 \times 190 = 570$ recesses in one year.

? **How can you use multiplication to describe other events in a school year?**

A. List 4 other school events that happen several times in a day, a week, or a month.

B. How many times does each event happen?

C. What 2 pieces of information do you need to determine how often each event happens in a year?

D. Calculate the total number of times the events happen. Use multiplication.

E. Explain why you chose your multiplication methods.

Get your tickets now!

Ferris Wheel   5 tickets
Pirate Bounce   3 tickets

Buy your tickets in bundles
of 10, 30, or 60!

NEL

Dividing Multi-Digit Numbers

GOALS

You will be able to

- relate division to multiplication
- divide 2-digit numbers by 1-digit numbers in different ways
- estimate quotients
- create and solve division problems
- solve problems by guessing and testing

What division problems can you create and solve using information in this photo?

Chapter 10

Getting Started

Planning a Play Day

Sergey's class is planning a Play Day.
36 students will participate.
Students will be divided into equal groups
for some events.

? How can you group students or equipment for Play Day events?

A. Sergey put 18 beanbags into 2 baskets for the relay race. He put the same number into each basket. How many are in each basket? Explain your thinking.

B. Why can you use division to describe the situation in Part A? Write the equation.

C. Lauren is planning a beanbag toss. How many teams of 6 can she make with the 36 students? Explain how you know.

D. Why can you use division to describe the situation in Part C? Write the equation.

E. Create and solve a division problem about a Play Day event for 36 students. Write a division equation that represents your problem.

What Do You Think?

Do you *agree* or *disagree* with each statement? Explain your thinking.

1. Division is the opposite of multiplication.

2. When you divide a number by 4, the answer is greater than it would be if you divided the number by 2.

3. You can divide one number by another number by multiplying or adding or subtracting.

Exploring Division

You will need
- a 100 chart
- counters

GOAL

Solve division problems using models.

Aneela and Julia are making kaleidoscopes in the Science Club.

Each kaleidoscope has a reflector made of 3 plastic rectangles. Aneela and Julia plan to use at least 70 and no more than 90 plastic rectangles. They don't want any left over.

? **How many reflectors can Aneela and Julia make so that no rectangles are left over?**

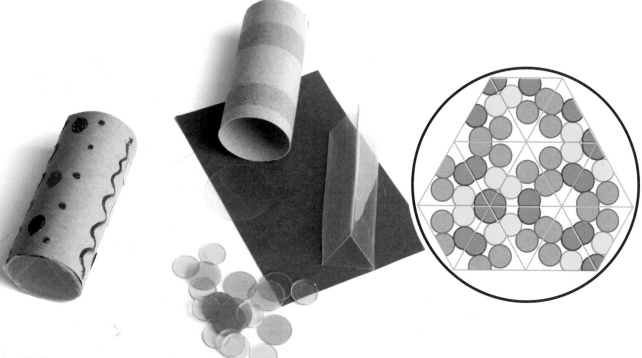

350

Relating Division to Multiplication

You will need
- a 100 chart
- counters

GOAL

Solve division problems by multiplying.

5 students are putting 90 spring rolls in packages of 6 to sell at the spring fair. They agreed to share the job fairly.

? **How can you multiply to decide how many packages of spring rolls each student should pack?**

Chapter 10
Lesson 3

Using Subtraction to Divide

You will need
- number lines

GOAL

Solve division problems by subtracting.

Kate is making decorations for a banquet.
She is tying balloons together in bunches of 4.
She has 90 balloons.

❓ How many balloons will be left over after Kate makes as many bunches of 4 balloons as she can?

Kate's Division

I'll use a number line to divide 90 by 4.
I'll keep subtracting bunches of 4 until there aren't
enough balloons to make any more bunches.

First I'll make 10 bunches. 10 is easy.
That uses 40 balloons.
There are 50 left.

10 groups of 4

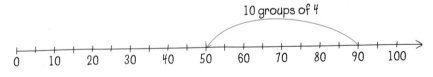

```
4)90
 -40 | 10
  50
 -40 | 10
  ▪
```

I'll do it again. That's 20 bunches so far.

10 groups of 4   10 groups of 4

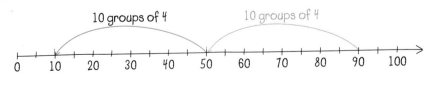

The amount left over after a number is divided into a whole number of equal parts

$$44 \div 7 = 6 \text{ R2}$$

quotient

6 R2
7) 44
remainder

After dividing, write the whole number part of the quotient with the remainder (R).

If the remainder is 0, write just the whole number part of the quotient.

A. Complete Kate's division to figure out the **quotient**.

B. How many bunches can Kate make?

C. What will the **remainder** be? Explain how you know.

Reflecting

D. Why do you think Kate started by subtracting 10 groups of 4?

E. How do you know she could have started by subtracting 20 groups of 4 instead?

Checking

1. Kate has 99 balloons to put in bunches of 6.
   a) How many bunches can she make?
   b) How many balloons will be left over?

Practising

2. Kate has 62 balloons to put in bunches of 5.
   a) How many bunches can she make?
   b) What will the remainder be?

3. Calculate.
   a) $3\overline{)42}$
   b) $6\overline{)62}$
   c) $50 \div 4 = \blacksquare$
   d) $\blacksquare = 96 \div 2$

4. How many pairs of moccasins are there?
   a) 26 moccasins
   b) 42 moccasins
   c) 92 moccasins

5. Ben and Anna each divided 57 by 5.
   a) Are both ways correct? How do you know?
   b) Whose way do you prefer? Why?

Ben

$$5\overline{)57}$$
$$\underline{-25}\;|5$$
$$32$$
$$\underline{-20}\;|4 \quad)\; 5+4+2=11$$
$$12$$
$$\underline{-10}\;|2$$
$$2$$

Anna

$$5\overline{)57}$$
$$\underline{-50}\;|10$$
$$7 \quad)\; 10+1=11$$
$$\underline{-5}\;|1$$
$$2$$

6. Olena cut out the dough for 51 *pyrohy* (perogies). She cut out 3 circles in each row of the dough. How many rows were there? Show your work.

7. **a)** Write a problem about balloons in bunches that can be solved by dividing 75 by 8.
   b) Solve your problem.

8. Jeff is dividing 73 by 5. He thinks his method is going to take too long. How can you do the same division in fewer steps?

$$5\overline{)73}$$
$$\underline{-5}\;|1$$
$$68$$
$$\underline{-5}\;|1$$
$$63$$
$$\underline{-5}\;|1$$
$$\blacksquare$$

9. Amélie and Grégoire are ordering juice for the *Soirée de la poésie* at their school. Juice boxes come in packages of 3. They need 81 juice boxes. How many packages do they need?

10. 2 different division calculations result in the same quotient. What might the calculations be? Use a 2-digit number divided by a 1-digit number.

11. You divide a number by 4 and the remainder is 3. How do you know the number is odd?

12. You divide a number by 5. The remainder is 2 more than the quotient. What could the number be? List 2 possible answers.

13. Tien, Joshua, and Cole showed how to divide 34 by 3 on a number line. They all said the answer was 11 R1. Explain how you think each student got the answer.
    a) Tien:

    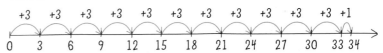

    b) Joshua:

    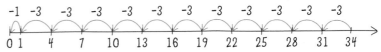

    c) Cole:

    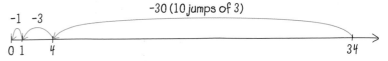

14. Explain how to divide 59 by 3 by subtracting groups.

Lesson 4 Dividing by Renaming

You will need
- base ten blocks

GOAL

Divide by renaming the dividend.

Lang and Aneela both collect postcards.
Lang has 3 times as many postcards as Aneela.

? **If Lang has 84 postcards, how many postcards does Aneela have?**

Joshua's Division

I have to calculate 84 ÷ 3.

I put the 84 in 3 rows of blocks to make it easier to divide by 3.

I can rename the **dividend** 84 as 60 + 24.

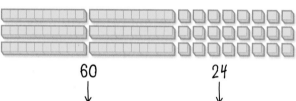

84 ÷ 3 is the same as 60 ÷ 3 added to 24 ÷ 3.

A. Why did Joshua divide 84 by 3?

B. How does the picture show that 84 ÷ 3 is the same as 60 ÷ 3 added to 24 ÷ 3?

C. Finish Joshua's division. How many postcards does Aneela have?

Reflecting

D. Why do you think Joshua renamed 84 as 60 + 24 instead of 80 + 4?

E. How can you use your answer to Part C to calculate 87 ÷ 4?

Checking

1. Tracy and Aaron both collect hockey cards. Tracy has 4 times as many cards as Aaron. Tracy has 92 cards.
   a) What calculation can you do to figure out the number of cards Aaron has?
   b) How can you rename 92 to solve the problem?
   c) How many cards does Aaron have? Show your work.

Practising

2. How can you rename the dividend to complete each division?
   a) 67 ÷ 3 = ▓
   b) 81 ÷ 5 = ▓
   c) ▓ = 57 ÷ 2
   d) ▓ = 75 ÷ 6

3. Divide by renaming. Explain your strategy for one division.
   a) 5)75
   b) 4)58
   c) 7)92
   d) 9)94

4. Joshua wanted to arrange 87 postcards equally in 3 albums. Read about his method at the left.
   a) Will Joshua's method work? Explain.
   b) Use his method to divide 72 by 4.

5. Show 2 ways to rename 78 to make it easier to divide it by 2. Explain how each way makes it easier to divide.

I can rename
87 as 90 − 3.
90 ÷ 3 = 30
3 ÷ 3 = 1
So I can put
30 − 1 = 29
postcards in
each album.

Frequently Asked Questions

Q: What are some ways to divide?

A1: You can think of the division as multiplication. For example, to calculate $48 \div 4$, think of what numbers to multiply.

$4 \times 10 = 40$ and $4 \times 2 = 8$

$48 \div 4 = 10 + 2$, so $48 \div 4 = 12$.

A2: You can subtract groups of the divisor from the dividend. For example, to calculate $77 \div 3$,

$$\begin{array}{r} 3\overline{)77} \\ -60 \\ \hline 17 \\ -15 \\ \hline 2 \end{array}$$

20
5
} 25

There are 25 groups of 3 and a remainder of 2.

A3: You can keep adding groups of the divisor until you reach, or almost reach, the dividend. For example, to calculate $47 \div 3$,

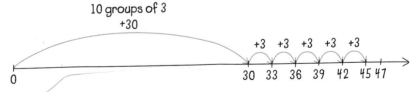

There are 15 groups of 3 in 45, so $47 \div 3 = 15$ R2.

A4: You can divide by renaming. Choose parts that are easy to divide by the divisor. For example, to calculate $66 \div 5$, you can rewrite 66 as $50 + 16$.

$50 \div 5 = 10$ and $16 \div 5 = 3$ R1

So, $66 \div 5 = 13$ R1.

Practice

Lesson 1

1. Melissa used tally marks to count the number of students in her school who have dogs. How many groups of 5 marks will there be if she counts 70 students? Show your work.

Lesson 2

2. What is the missing number in each calculation?
   a) $4 \times \blacksquare = 44$
   b) $3 \times \blacksquare = 72$
   c) $\blacksquare \times 22 = 88$
   d) $\blacksquare \times 15 = 45$

Lesson 3

3. Tony has 83 jars of jam to pack in boxes of 6. How many boxes can he fill? How many are left over?

4. Divide.
   a) $5\overline{)71}$
   b) $82 \div 3 = \blacksquare$
   c) $7\overline{)82}$
   d) $\blacksquare = 97 \div 4$

5. Frogs are decorative closures used on jackets. Kathy sews 4 frogs on each jacket. How many jackets can she finish with 65 frogs?

6. How much less is $56 \div 4$ than $57 \div 3$?

Lesson 4

7. How can you use $78 = 60 + 18$ to calculate $78 \div 6$?

8. Kativa has attended 4 times as many Kathaka dance classes as Monika. Kativa has gone to 68 classes. How many classes has Monika attended?

9. Explain why $78 \div 3$ is the same as $60 \div 3$ added to $18 \div 3$.

Lesson 5

Estimating Quotients

GOAL

Use multiplication and division facts to estimate quotients.

Julia made a list of the books she would like to read.

Julia's Reading Plan

I want to read *Sarah, Plain and Tall* in 3 days.
I'll read about the same number of pages every night.

Books to Read

Title	Author	Pages
The Girl Who Loved Wild Horses	Paul Goble	31
The Missing Sun	Peter Eyvindson	47
Earthlings Inside and Out	Valerie Wyatt	63
Jacob Two-Two and the Dinosaur	Mordecai Richler	85
Sarah, Plain and Tall	Patricia MacLachlin	58
Esio Trot	Roald Dahl	62
Cam Jansen and the Mystery of Flight 54	David A. Adler	56

? **About how many pages will Julia read each night?**

A. How do you know that Julia will read more than 10 pages each night?

B. How do you know that Julia won't read more than 20 pages each night?

C. How would dividing 60 by 3 help Julia estimate the number of pages she has to read each night?

D. How can Julia use $6 \div 3 = \blacksquare$ to calculate $60 \div 3$?

E. About how many pages will Julia read each night?

Reflecting

F. Why might it be useful for Julia to think of 58 as about 6 tens?

G. Suppose Julia decides to take 5 days to read the book. How might she estimate the number of pages to read each day?

Checking

1. Julia wants to read *Esio Trot* in 5 days. She plans to read about the same number of pages each day.
   a) How do you know she will read more than 10 pages each day?
   b) About how many pages will she read each day? Explain.

Practising

2. Julia wants to read the first 2 books on her list in 4 days. She plans to read about the same number of pages each day.
   a) About how many pages will she read each day?
   b) About how many pages would she have to read each day to finish in 8 days instead of 4?
   c) How can your answer to part (a) help you answer part (b)?

3. Estimate each quotient. Explain your thinking.
   a) $6\overline{)92}$
   b) $91 \div 8$
   c) $7\overline{)89}$
   d) $47 \div 3$

4. Estimate to decide which quotient is closest to 10.
   A. $92 \div 6$ **B.** $71 \div 3$ **C.** $68 \div 5$ **D.** $83 \div 9$

5. a) 9 students want to raise $96 to buy a used koto for the school's Japanese music night. Why might you estimate $96 \div 9$ as about 10 to figure out about how much each student needs to raise?
   b) Suppose the students had to raise only $76. About how much would each student have to raise? How do you know?

Reading Strategy

Read the problem. Predict what you need to do to solve the problem. Share your prediction with a classmate.

6. Gerry can finish his book if he reads about 30 pages every day for 3 days. About how long is the book?

7. ▉ $\div 4$ is between 10 and 20. What numbers might ▉ be?

8. About how much less is $49 \div 2$ than $68 \div 2$? How do you know?

9. Estimate $77 \div 2$ at least 2 different ways. Explain your thinking.

3 Card Quotient

Number of players: 2 to 4

How to play: Arrange 3 cards to make a division calculation with the greatest quotient.

You will need
- number cards 0 to 9 (4 of each)

- **Step 1** Give 3 cards to each player.

- **Step 2** All players arrange their cards to make a 2-digit number divided by a 1-digit number.

- **Step 3** All players calculate their quotients. The player with the greatest quotient wins 1 point.

- **Step 4** Use the leftover cards to start the next round.

Continue playing until one player has 5 points.

Tien's Division

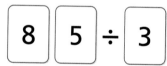

When I divide 85 by 3, the quotient is 28 R1.

Lang played $86 ÷ 4 = 21$ R2 and Cole played $96 ÷ 6 = 16$.

I have the greatest quotient, so I get a point.

Lesson 6

Dividing by Sharing

You will need
• base ten blocks

GOAL

Solve division problems using base ten blocks.

Cole had 94 fishing lures. He plans to give them to his 4 uncles. He wants to give the same number to each uncle.

? How many fishing lures will each uncle get?

Cole's Division

I think each uncle will get about 25 lures because $4 \times 25 = 100$.

To get the exact number, I'll divide 94 by 4 using the base ten blocks.

• **Step 1** I'll model 94 lures.

$$4\overline{)94}$$

• **Step 2** I'll divide the tens into 4 groups. There are 2 tens in each group and 1 ten left over, so I write 20 in the quotient. I've used 80 lures and there are 14 left to share.

$$\begin{array}{r} 20 \\ 4\overline{)94} \\ -80 \\ \hline 14 \end{array}$$

- **Step 3** I can regroup the ten as 10 ones.
  I'll put the 14 ones into the 4 groups.
  I can put 3 ones into each of the 4 groups.
  I record the 3 in the quotient.
  The quotient is now 20 + 3 = 23.
  The remainder is 2.

$$\begin{array}{r} 3 \\ 20 \\ 4\overline{)94} \\ -80 \\ \hline 14 \\ -12 \\ \hline 2 \end{array}$$

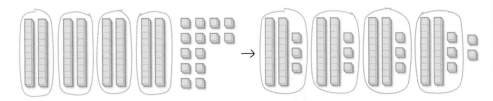

Each uncle will get 23 lures. There are 2 lures left over.
This answer is reasonable because 23 is close to my estimate
of 25. I'll check by multiplying and adding the remainder.
23 × 4 = 92 and 92 + 2 = 94. The division is correct.

Reflecting

A. How is this method like other division methods you know? How is it different?

B. Why could Cole check his answer by multiplying and adding the remainder?

Checking

1. The people in 6 fishing boats caught 72 fish. Each boat caught the same number of fish.
   a) How many fish did each boat catch?
   b) How would the answer change if there had been 75 fish?

Practising

2. Calculate. Show your work.
   a) $5\overline{)74}$ b) $6\overline{)84}$ c) $4\overline{)98}$ d) $7\overline{)84}$

3. Calculate, then check your answers.
   a) $78 \div 6 = \blacksquare$ c) $\blacksquare = 93 \div 7$
   b) $95 \div 3 = \blacksquare$ d) $\blacksquare = 65 \div 4$

4. The video store has 85 cartoon DVDs for rent. They are displayed in 5 sections with about the same number of DVDs in each section.
   a) What equation can you use to figure out how many DVDs are in each section?
   b) Solve your equation.
   c) Is your answer reasonable? How do you know?

5. Some students want to play the Aboriginal game Pine Cone Hoop Toss, where pine cones are tossed into a hoop on the ground. They have 95 pine cones. Each student needs 6 pine cones. How many students can play? Show your work.

6. Rory and her brother have 54 new stamps to share equally between them.
   a) How many stamps should each get?
   b) If they decide to share equally with their cousin, how many will each get? Show your work.

7. Show one way to complete this division so that no digit is used more than once. Explain your strategy.
   $\blacksquare\blacksquare \div 4 = 1\blacksquare$ R2

8. The distance all the way around a hexagon with all sides equal is 84 cm. How long is each side?

9. When you divide using base ten blocks, why is it important to have the same number of tens in each group? Explain using $92 \div 4$.

Remainder Magic

Try this curious number trick.

- Ask a classmate to think of a number less than 30. Ask him or her to divide that number by 2, by 3, and by 5, and to tell you only the remainder each time.
- Multiply the remainder for 2 by 15.
- Multiply the remainder for 3 by 10.
- Multiply the remainder for 5 by 6.
- Add the 3 products.

If the sum is less than 30, you have the number.
If it is not, then subtract 30 from the sum to get the number.

Joshua's Number

I picked the number 21.
21 ÷ 2 has a remainder of 1 and 1 × 15 = 15.
21 ÷ 3 has a remainder of 0 and 0 × 10 = 0.
21 ÷ 5 has a remainder of 1 and 1 × 6 = 6.

15 + 0 + 6 = 21

Kate's Number

I picked the number 17.
17 ÷ 2 has a remainder of 1 and 1 × 15 = 15.
17 ÷ 3 has a remainder of 2 and 2 × 10 = 20.
17 ÷ 5 has a remainder of 2 and 2 × 6 = 12.

15 + 20 + 12 = 47
47 − 30 = 17

1. Try the trick with a classmate using several different starting numbers.

Solving Problems by Guessing and Testing

GOAL

Use guessing and testing to solve problems.

Aneela's older brother is learning to use a saw to cut wood carefully. He has 3 pieces of wood to practise on. They are all the same length and he knows they are all longer than 40 cm.

If he cuts one into pieces 3 cm long, 1 cm is left over.

If he cuts one into pieces 4 cm long, 1 cm is left over.

If he cuts one into pieces 5 cm long, 1 cm is left over.

 How long might the pieces of wood be?

Aneela's Solution

Understand
I need a number that leaves a remainder of 1 when I divide by 3 or 4 or 5. The length is greater than 40 cm.

Make a Plan
I'll try 41 first. I was going to try 40, but then there wouldn't be a remainder of 1 when I divide by 4. I'll try other numbers that end in 1 if 41 doesn't work.

Carry Out the Plan

41 ÷ 5 = 8 R1
41 ÷ 4 = 10 R1
41 ÷ 3 = 13 R2   It's not 41.

I'll try 51.
51 ÷ 5 = 10 R1
51 ÷ 4 = 12 R3   It's not 51.

I'll try 61.
61 ÷ 5 = 12 R1
61 ÷ 4 = 15 R1
61 ÷ 3 = 20 R1   61 worked.

The pieces of wood could be 61 cm long.

Reflecting

A. Do you think it was a good idea to test numbers that end in 1? Explain.

B. Why was guess and test a good strategy for Aneela to choose? Explain.

Checking

1. Shilpa was putting samosas into bags. When she put them into bags of 6, she had 1 left over. When she put them into bags of 8, she had 1 left over. How many samosas might she have? Give 2 possible solutions.

Practising

2. A number is divided by 3. The remainder is 1. When the quotient without the remainder is divided by 5, the remainder is 2. What might the original number be?

3. Use only odd numbers for all of the missing digits to make this division calculation true.

4. Massimo had between 50 and 60 square tiles. He made a rectangle that was 6 tiles long. He had 4 tiles left over. What might Massimo's rectangle look like? Sketch it. Label the side lengths.

5. Jeff chose a number. He added it to the next 2 counting numbers greater than it. The sum was 201. What are the 3 numbers?

6. You multiply a number by 3 and then divide it by 2, and the answer is 12. What is the number?

7. Annie bought 2 big pails and 1 small pail. The total cost was $60. The small pail cost half as much as a big one. How much did the small pail cost?

8. Derek and Rain shared some stickers equally. Then they decided to share their stickers equally with Terry. They each give Terry 10 stickers. How many stickers are there?

9. The area of a rectangle is 72 square units. The length is 14 more units than the width. What is the length?

10. Create your own guessing and testing problem. Give your problem to a classmate to solve.

Remainder Hunt

Number of players: 2 to 5

How to play: Arrange 3 cards to make a division calculation with the least possible remainder.

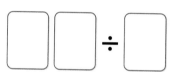

- Step 1 Each player gets 3 cards.

- Step 2 All players arrange their cards to make a 2-digit number divided by a 1-digit number. Try to make the remainder as close to 1 as possible.

- Step 3 All players calculate their quotients, with remainders. The player with the least remainder wins 1 point.

Continue playing until one player has 5 points.

Cole's Division

$23 \div 5$ has a remainder of 3.
$25 \div 3$ has a remainder of 1.
I'll use $25 \div 3$.

3 5 2

Tien has $24 \div 6$, which has a remainder of 0.

Lang has $16 \div 7$, which has a remainder of 2.

Tien gets a point.

Chapter Review

Frequently Asked Questions

Q: How can you estimate a quotient?

A: You can use a number close to the dividend that is easy to divide. For example, to estimate $97 \div 2$, 97 is about 100, and 100 is 10 tens.
10 tens $\div$ 2 is 5 tens, or 50.

Q: How can you divide by sharing?

A: You can use models such as base ten blocks and make sure each group gets the same number. The quotient is the number in each group. For example, $66 \div 5$:
You need to make 5 groups. Put a ten in each group. Record 10 in the quotient of the calculation.

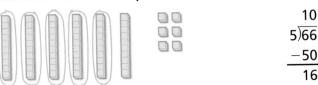

$$
\begin{array}{r}
10 \\
5\overline{)66} \\
-50 \\
\hline
16
\end{array}
$$

There are 16 left over, which you can regroup as ones. Place the ones into the 5 groups.

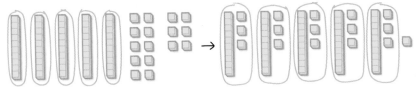

Each group now has $10 + 3 = 13$.
The remainder is 1.
Check using multiplication and addition:
$5 \times 13 = 65$ and $65 + 1 = 66$.
The division is correct.

$$
\begin{array}{r}
3 \\
10 \\
5\overline{)66} \\
-50 \\
\hline
16 \\
-15 \\
\hline
1
\end{array}
$$

Practice

Lesson 2

1. How can you use multiplication to solve 64 ÷ 2?

Lesson 3

2. Kyle is making treat bags with 3 stickers in each bag. If he has 53 stickers, how many treat bags can he make? How many stickers are left over?

3. a) How many jumps of 5 can you make on a number line like this? Show your work.

0   63

   b) What calculation did you represent?

4. Which calculation has the greatest remainder?
   A. 57 ÷ 4   B. 62 ÷ 5   C. 73 ÷ 6   D. 85 ÷ 7

Lesson 4

5. How can you use 64 = 50 + 10 + 4 to calculate 64 ÷ 5?

6. Adrian has 4 times as many stamps as Sunita. If Adrian has 76 stamps, how many does Sunita have?

Lesson 5

7. Which quotients are about 20? Explain how you estimated.
   A. $4\overline{)79}$ B. $3\overline{)46}$ C. $7\overline{)89}$ D. $5\overline{)96}$

8. Celia makes 3 bead bracelets a day. About how many days will it take her to make 70 bead bracelets?

9. There are 52 weeks in a year. About how many weeks are in each of the 4 seasons?

Lesson 6

10. Calculate. Check your answers.
    a) 76 ÷ 6 = ▨   c) ▨ = 54 ÷ 4
    b) 99 ÷ 8 = ▨   d) ▨ = 89 ÷ 3

11. 4 people hold up each dragon. How many dragons can 50 people hold up?

12. 5 basketball players shared the $65 cost of the coach's gift equally. How much did each player pay?

Lesson 7

13. Mia always had a marble left over whether she bagged her marbles in 2s, 3s, or 4s. How many marbles might she have had? Give 2 answers.

14. Hannah divided a 2-digit number by 8. The remainder was more than the whole number part of the quotient. What could the number have been? Explain your thinking.

Lesson 8

15. Use the same digit in both ▨ to make this equation true: 9▨ ÷ ▨ = 23 R2.

What Do You Think Now?

Look back at **What Do You Think?** on page 349. How have your answers and explanations changed?

374

Chapter Task

Heritage Poster Prizes

Sponsors donated 95 music downloads for the
top 8 winners in the heritage poster project.
The top winner got an extra 7 downloads.
All the other winners got the same number.

The farmhouse my great-grandfather
built when he moved to Canada

**❓ How many downloads could each
winner have received?**

A. How many downloads did each winner get?
   Explain and check your answer.

B. Suppose the prizes only went to the top 6 winners
   and the top winner gets an extra 5 downloads. How
   many downloads would each winner get? Explain.

C. How many people (up to 10) could be winners if
   each winner gets exactly the same prize? How many
   downloads would they win? Explain your answer.

Chapter 11

3-D Geometry

GOALS

You will be able to

- identify and describe prisms in the environment
- compare and describe triangular and rectangular prisms
- sort prisms
- construct triangular and rectangular prisms

How can you describe the shape of this longhouse using geometry words?

Getting Started

You will need
- 3-D models

Describing Packages

These students are wrapping packages of different sizes and shapes.

? **How can you describe the sizes and shapes of the packages?**

Geometry Word List
face   vertex edge
square quadrilateral
rectangle pentagon cone
hexagon octagon
cylinder sphere triangle
          pyramid
          prism

A. Examine 2 packages in the picture. Make a list of geometry words that describe each package.

B. Compare the 2 packages. How are they the same? How are they different?

C. Name 2 things that might come in each package.

D. Choose one package. How many **edges** and **vertices** does it have?

What Do You Think?

Do you *agree* or *disagree* with each statement? Explain your thinking.

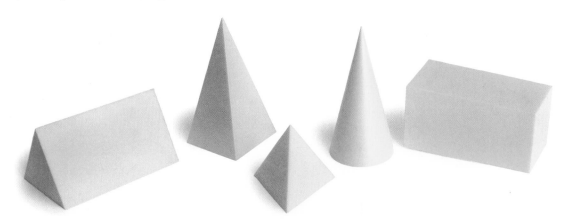

1. 3-D objects always have more **vertices** than **faces**.

2. It is possible for a 3-D object to have only **rectangular faces**.

3. It is possible for a 3-D object to have only **triangular faces**.

4. It is not possible for a 3-D object to have only one triangular face.

Recognizing Rectangular Prisms

You will need
- storage boxes
- rectangular prisms

> **GOAL**
>
> Explore the attributes of rectangular prisms.

prism
(right prism)

A 3-D object with opposite faces that are **polygons** of the same size and shape, and other faces that are rectangles

These are prisms.

These are not prisms.

rectangular prism

A prism that has all rectangular faces

Hailey's family is moving. They use boxes shaped like **rectangular prisms** to pack their things.

(?) **How are these packing boxes the same or different?**

380

Making Rectangular Prisms with Cubes

You will need
- linking cubes

Alec's Rectangular Prisms

I can build 2 different rectangular prisms with 4 linking cubes.

I can build only 1 rectangular prism with 5 linking cubes.

I can also build some L shapes with 3, 4, or 5 linking cubes. They are prisms, but not rectangular prisms.

1. Try to build rectangular prisms with an odd number of cubes. What do you notice?

2. Try to build rectangular prisms with different numbers of cubes up to 20. Which number makes the most prisms?

Recognizing Triangular Prisms

You will need
- triangular prisms

GOAL

Describe the attributes of triangular prisms.

Diane went to a stunt show where skateboarders performed on ramps.

Diane made models of some ramps using modelling clay.

 What is the same about all the ramps?

Diane's Models

The models all have 5 faces, 6 vertices, and 9 edges. Each prism has 2 opposite faces the same size that are triangles. These are the **bases**. That's why these are all called **triangular prisms**.

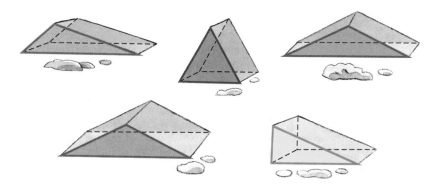

The other 3 faces of each of these triangular prisms are rectangles. The rectangular faces can be all the same size or different sizes.

bases

The 2 polygon faces that are the same size and shape in a prism

bases

bases bases

triangular prism

A prism with triangles as bases

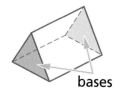

bases

Reflecting

A. How are these triangular prisms different?

B. How are all triangular prisms the same?

<table>
</table>

Communication Tip

The bases of a prism are rectangles only in rectangular prisms. Any pair of opposite rectangular faces can be the bases.

In other prisms, other polygon shapes can be the bases.

Reading Strategy

Read the problem. Sketch a picture for each part to help you solve the problem.

Checking

1. How do you know that this 3-D object is a triangular prism?

Practising

2. Which 3-D object is a triangular prism? Explain.

A   B   C   D   E

3. Which of these are possible? Explain how you know.
   a) a triangular prism that has 3 rectangular faces the same size
   b) a triangular prism that has only 2 of the 3 rectangular faces the same size
   c) a triangular prism that has 3 rectangular faces that are different sizes

4. a) How are these prisms alike?
   b) How are these prisms different?

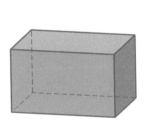 

MATH GAME

Shape Match

Number of players: 2 or more

How to play: Match the clue cards to the prisms.

You will need

- Shape Match Clue Cards (2 sets) (blackline master)
- a die
- 3-D models of prisms

- **Step 1** Shuffle the clue cards. Place them face down in a pile.

- **Step 2** The first player draws a clue card from the pile and rolls the die. The number rolled goes in the blank on the card.

- **Step 3** If the description on the clue card matches a prism, the player takes the prism. If there is no prism to match the description, the player gets nothing.

- **Step 4** Take turns drawing cards, rolling, and collecting prisms if they match.

Continue playing until one player has collected 4 prisms.

Michael's Turn

I picked a clue card and rolled the die.

> I have more than 5 faces.

I need to find a 3-D shape with more than 5 faces.

I'll pick a prism with a hexagon base.

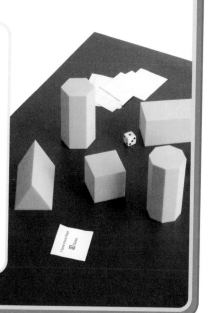

Communicating about Prisms

You will need
- 3-D Objects (blackline master)

GOAL

Identify prisms in the environment and sort them.

For a project, Annie's group is designing some toys and furniture for a daycare centre. She describes her ideas in an e-mail message to the rest of her group.

? How can you describe and sort objects by their shapes?

Annie's Description

New playhouse for daycare

To: Michael, Alec, Diane
Subject: New playhouse for daycare

Hi,
I think our plan should include a playhouse. It could be pretty cool. The top part of the playhouse has a triangular face. The bottom part is a prism. Can you picture it?
Annie

A. Improve on Annie's description of the top part of the playhouse. Use the Communication Checklist.

B. Improve on Annie's description of the bottom part of the playhouse. Use the Communication Checklist.

C. Describe the other objects Annie imagined.

D. Write the name of each object on a piece of paper and sort them. What attributes did you use to sort them?

Reflecting

E. What information do you need to identify a prism?

Checking

1. How can you sort the objects Annie imagined in different ways? Name the attribute you used each time.

Reading Strategy

Brainstorm a list of words that will help you describe a triangular prism.

Practising

2. a) Find a prism in the classroom and describe it to a classmate.
   b) Can your classmate identify your prism from your description? If not, how can you improve it?

3. Describe another prism in the classroom. Use the Communication Checklist.

4. a) Sort these 3-D objects.
   b) Explain how you sorted the objects. Use the Communication Checklist.

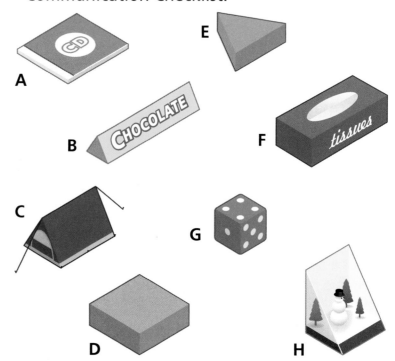

5. a) Finish Diane's description of the pencil case.
   b) Check your work using the Communication Checklist.

6. Do you think there are more rectangular or more triangular prisms in the environment? Explain your answer.

7. What are the most important things to think about when describing a 3-D object?

Constructing Prisms

You will need
- pattern blocks
- modelling clay
- dental floss

GOAL

Construct prisms using pattern blocks and modelling clay.

Ken is watching his sister play with a toy puzzle. He notices that the objects are prisms. Ken wants to make some of these prisms using pattern blocks and modelling clay.

? **What different prisms can you make with these materials?**

Lesson 5

Constructing Prisms from Nets

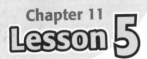

You will need
- Prism Nets (blackline master)
- tape

net

A 2-D pattern that can be folded into a 3-D object

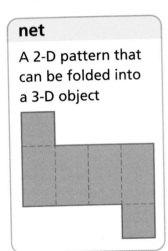

Construct prisms using 2-D nets.

Alec and Hailey found these **nets** while sorting boxes in the recycling bin.

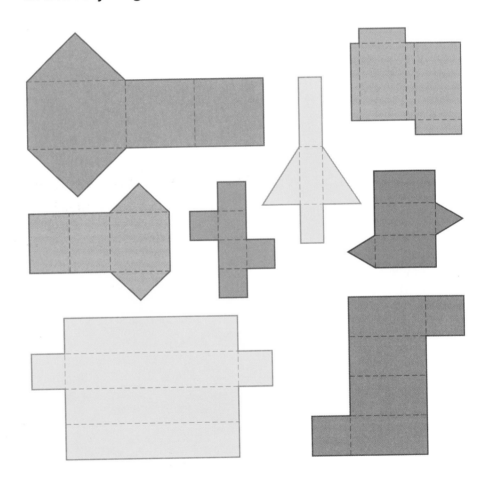

 How can you make prisms using nets?

Chapter Review

Frequently Asked Questions

Q: **How are all prisms the same?**

A: All prisms (right prisms) are 3-D objects. They all have at least 2 polygon faces (the bases) that are the same size and shape on opposite sides. All other faces on a prism are rectangles that touch both bases.

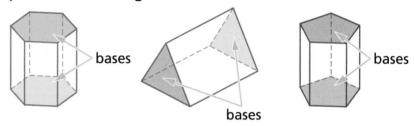

bases

bases

bases

Q: **How can you tell the difference between a rectangular prism and a triangular prism?**

A: All of the faces of a rectangular prism are rectangles. A rectangular prism has 6 faces, 12 edges, and 8 vertices.

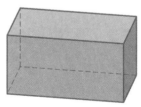

2 of the faces of a triangular prism are triangles and the other 3 faces are rectangles. A triangular prism has 5 faces, 9 edges, and 6 vertices.

Practice

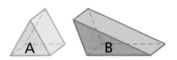

Lesson 2

1. **a)** How are the prisms at the left alike?
    b) How are the prisms different?

Lesson 3

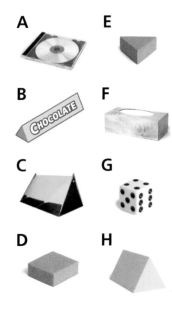

2. **a)** Sort these objects according to their shapes.
    b) Explain how you sorted the objects.
    c) Describe the shape of one of the objects using geometry language. Do not name the object.
    d) Show your description to a classmate. Can your classmate identify your prism using your description? If not, how can you improve it?

Lesson 4

3. **a)** Construct a model for prism C in Question 2.
    b) Construct a model for prism F in Question 2 using a different method.

Lesson 5

4. **a)** What kind of prism will each net below make?
    b) Cut out nets like these. Make the prisms and name them.

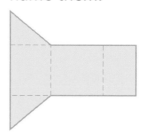
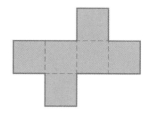

What Do You Think Now?

Look back at **What Do You Think?** on page 379. How have your answers and explanations changed?

Chapter Task

Making a 3-D Photo Holder

Ken wants to display his photos on a 3-D photo holder. He wants to pack the holder into a box to mail it to his grandparents.

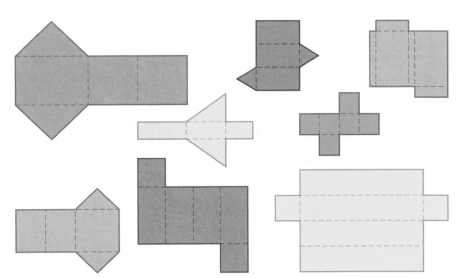

❓ **How can Ken make a photo holder and a packing box?**

A. Make prisms out of Ken's nets.

B. Choose one prism to be the photo holder. Explain your choice.

C. Choose another prism to be the box to mail the holder in. Explain your choice.

D. Can you choose any other 2 prisms to be the photo holder and packing box? Why or why not?

1. What time does the clock show?
   - A. 9:13 a.m.
   - C. 1:09 a.m.
   - B. 3:09 p.m.
   - D. 1:09 p.m.

2. Which date shows March 21, 2008?
   - A. 2008-03-21
   - C. 01-21-2008
   - B. 2008-03-08
   - D. 2008-04-21

3. Jon checked the clock to see if it was time for lunch. What is the time?
   - A. 11:25 a.m.
   - C. 12:30 a.m.
   - B. 11:35 a.m.
   - D. 12:25 p.m.

4. Determine the area of this octagon.
   - A. 28 square units
   - B. 20 square units
   - C. 7 square units
   - D. 32 square units

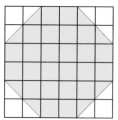

5. Identify the expression with the same product as 46 × 2.
   - A. 23 × 4   B. 94 × 1   C. 23 × 1   D. 48 × 4

6. A puzzle box contains 120 pieces. About how many pieces would be in 8 puzzles the same size?
   - A. more than 800, but fewer than 1600
   - B. more than 80, but fewer than 800
   - C. more than 1600
   - D. fewer than 800

7. 92 slices of cake were sold at the fun fair. Each cake was cut into 8 slices. How would you calculate the number of cakes that were sold?
   A. Multiply 92 by 8.
   B. Divide 92 by 8.
   C. Add 92 and 8.
   D. Subtract 8 from 92.

8. Identify the expression with the same quotient as 72 ÷ 3.
   A. 36 ÷ 3   B. 36 ÷ 6   C. 96 ÷ 4   D. 48 ÷ 4

9. One ride at an amusement park holds 4 people in each seat. 62 people take the ride. How many seats are needed?
   A. 14   B. 15   C. 16   D. 17

10. What is the name of this 3-D object?
   A. cube
   B. cylinder
   C. rectangular prism
   D. triangular prism

11. What kind of 3-D object will this net make?
   A. pyramid
   B. cylinder
   C. rectangular prism
   D. triangular prism

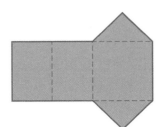

Answers

Chapter 1
Patterns in Mathematics

Lesson 1: pages 4–6
Patterns in an Addition Table

1.

+	2	4	6	8	10	12	14
2	4	6	8	10	12	14	16
4	6	8	10	12	14	16	18
6	8	10	12	14	16	18	20
8	10	12	14	16	18	20	22
10	12	14	16	18	20	22	24
12	14	16	18	20	22	24	26
14	16	18	20	22	24	26	28

2.

+	2	4	6	8	10	12	14
1	3	5	7	9	11	13	15
3	5	7	9	11	13	15	17
5	7	9	11	13	15	17	19
7	9	11	13	15	17	19	21
9	11	13	15	17	19	21	23
11	13	15	17	19	21	23	25
13	15	17	19	21	23	25	27

3.

+	10	20	30	40	50	60	70
1	11	21	31	41	51	61	71
2	12	22	32	42	52	62	72
3	13	23	33	43	53	63	73
4	14	24	34	44	54	64	74
5	15	25	35	45	55	65	75
6	16	26	36	46	56	66	76
7	17	27	37	47	57	67	77

4. e.g., Agree. In each row, if you go from left to right it is an increasing pattern, and if you go from right to left it is a decreasing pattern. For example, in the row starting with 5, the pattern going from left to right is an increasing pattern that adds 10 each time. The pattern going from right to left is a decreasing pattern that subtracts 10 each time.

Lesson 2: pages 8–11
Extending Patterns in Tables

1. a) yes
   b) e.g., Start with 4 and add 4 each time.

2. a) 35 rocks
   b) e.g., The pattern starts at 5 and adds 5 in each new row as it goes down the column.

3. a) 25 squares
   b) e.g., Start with 5 and add 4 each time.

4. a) 24 toothpicks
   b) e.g., Start with 3 and first add 4, then 3, then 4, then 3, and so on.
   c) 10

5. 6 shapes

6. a) e.g., My inuksuk is made of 2 large rocks and 5 small rocks.
   b) e.g., 12 large rocks, 30 small rocks
   c) e.g., no

Lesson 3: pages 12–14
Representing Patterns

1. a)

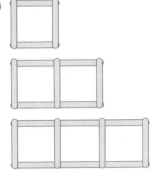

b) e.g., Each new shape has 3 more sticks than the one before.
c) 9 squares

2. a) the numbers 19, 16, and 13 modelled with base ten blocks
b) e.g., The number of chairs goes down by 3 each time.
c) 6 times

3. a) e.g., the numbers 10, 25, 40, and 55 modelled with base ten blocks
b) e.g., At each step, 1 more ten and 5 more ones are added. When you get 10 ones, you can trade them in for 1 ten.
c) 100 sit-ups

4. a) e.g., Bree's model has one frame with 6 sides. Bree's table shows the number of sticks going up by 5 each time, except after 15 it goes up by 6.
b) e.g., It was easier to see in the model because I could just see it, and I didn't have to count the sides.

5. a) e.g.,

1	19
2	18
3	16
4	13
5	9

b) the numbers 19, 18, 16, 13, and 9 modelled with base ten blocks
c) e.g., They both show a decreasing pattern starting with 19. The difference between the numbers increases by 1 each time. They are different because the table just gives numbers, while the model shows a concrete representation of the numbers.

Mid-Chapter (1) Review pages 16–17

1.

+	1	2	3	4	5	6	7
11	12	13	14	15	16	17	18
22	23	24	25	26	27	28	29
33	34	35	36	37	38	39	40
44	45	46	47	48	49	50	51

2. a) 13
b) e.g., Start with 3 and add 2 each time.

3. a) e.g., the numbers 12, 24, 36, 48, and 60 modelled with base ten blocks
b) 96

Lesson 4: pages 18–19
Solving Problems Using Patterns

1. 10 clowns

2. a) 6 clowns
b) 3 clowns

3. 2 times

4. Sunday

5. e.g., Brenna walks to the store every 2nd day and to the park every 5th day. How many days each month does she walk both places? (Answer: 3)

Lesson 5: pages 20–23
Solving Equations

1. a) e.g., $77 - \blacksquare = 73$ b) 4

2. a) 7 people

3. a) 2 b) 3 c) 4 d) 5

4. a) e.g., $11 + \blacksquare = 26$ b) $15

5. a) 9 b) 16 c) 44 d) 9 e) 16
   f) 16

6. a) e.g., $59 - \blacksquare = 54; 54 - \blacksquare = 52;$
   $52 - \blacksquare = 47; 47 - \blacksquare = 45;$
   $45 - \blacksquare = 40; 40 - \blacksquare = 38$
   b) 5, 2, 5, 2, 5, 2
   c) e.g., The number in the pattern decreases by 5, then by 2, then by 5, then by 2, and so on.

7. a) e.g., When you add something to 6, you get 45. b) 39

Lesson 6: pages 24–26
Solving Problems with Equations

1. a) 132 g b) e.g., guessing and testing

2. 28 kg

3. 62 flyers

4. 39 raffle tickets

5. 57 corn plants

6. 115 beads

7. a) e.g., Victor has already read 92 pages of his new book. The book has 137 pages. How many more pages does Victor have to read to finish the book?
   b) e.g., Alicia's school was given 78 bowling passes to give away to students. By lunchtime, there were only 23 passes left. How many bowling passes had already been given out?

Chapter 1 Review pages 28–30

1.

+	2	4	6	8	10
10	12	14	16	18	20
11	13	15	17	19	21
12	14	16	18	20	22
13	15	17	19	21	23

2. a) 50 toothpicks
   b) e.g., This is the pattern rule: Start with 4, then add 6, then add 4, and so on.

3. a) e.g., the numbers 8, 16, 24, and 32 modelled with base ten blocks
   b) 56 people

4. 2 times

5. a) 13 b) 32 c) 15 d) 7

6. a) $16 + \blacksquare = 32$ b) $16

7. $47

8. 244 pennies

Chapter 2
Numeration

Lesson 1: pages 36–37
Modelling Thousands

1. 42 hundreds

2. a) e.g., 7 thousands
   b) e.g., 2 thousands

3. a) 4 thousands 1 hundred
   b) 1 thousand 7 hundreds

4. a) 90
   b) 900

5. a) 20 hundreds   **b)** e.g., You might want to describe 2000 as 20 hundreds if you were comparing 2000 to another number that was given in hundreds.

Lesson 2: pages 38–41
Place Value

1. a) sketch of 19 hundreds, 4 tens, 3 ones
   b) sketch of 1 thousand, 9 hundreds, 4 tens, 3 ones

2. a) sketch of 1 thousand, 8 hundreds, 7 tens, 3 ones
   b) sketch of 6 thousands, 3 tens, 7 ones
   c) sketch of 4 thousands
   d) sketch of 6 thousands, 2 hundreds, 1 ten

3. e.g., sketch of 2 thousands, 6 tens, 5 ones; sketch of 1 thousand, 10 hundreds, 6 tens, 5 ones

4. a) 7126   **b)** 1010   **c)** 1742   **d)** 2718
   e) 1006

5. a) sketch of 1 thousand, 8 hundreds, 3 ones
   b) sketch of 10 hundreds, 2 ones
   c) sketch of 10 hundreds, 2 tens
   d) sketch of 1 thousand, 10 hundreds, 1 one

6. a) sketch of 4 thousands, 5 hundreds, 6 tens, 2 ones
   b) first day: sketch of 4 thousands, 6 hundreds, 6 tens, 2 ones
   second day: sketch of 4 thousands, 7 hundreds, 6 tens, 2 ones
   third day: sketch of 4 thousands, 8 hundreds, 6 tens, 2 ones
   fourth day: sketch of 4 thousands, 9 hundreds, 6 tens, 2 ones
   fifth day: sketch of 5 thousands, 6 tens, 2 ones

c) 5062 spring rolls sold
d) e.g., the number of hundreds changed

7. a) 6200
   b) e.g., 5900, 6000, 6400

8. e.g., 1010, 1040, 1060

9. a) sketch of 2 thousands, 2 hundreds, 2 tens, 2 ones
   b) 2 thousands, 2 hundreds, 2 tens, 2 ones

10. a) e.g., 9411
   b) e.g., 8223, 2940, 4542, 1185

11. e.g., I need at least 4 digits because if I add 1 to 999, my number is 1000. All numbers greater than 1000 have 4 digits.

Lesson 3: pages 42–44
Expanded Form

1. a) four thousand five hundred sixty-seven
   c) 4 thousands, 5 hundreds, 6 tens, 7 ones
   d) 4000 + 500 + 60 + 7

2. a) 8 thousands + 1 ten + 7 ones
   b) 8017

3. a) 2050; 2000 + 50
   b) 3014; 3000 + 10 + 4

4. a) e.g., 2970, 6651
   b) e.g., 2 thousands + 9 hundreds + 7 tens; 6 thousands + 6 hundreds + 5 tens + 1 one

5. a) 6066   **b)** 1210

6. 100

7. e.g., First, I look to see how many thousands there are. I put this digit in the thousands place. In the example, this digit is 3. Next, I look for hundreds, but there aren't any. I put 0 in the hundreds place. Next, I look for tens. Since 20 is 2 tens, I put 2 in the tens place. Then I put the ones digit in the ones place. This digit is 4. The standard form is 3024.

Mid-Chapter (2) Review pages 46–47

1. a) 4300 candies **b)** 430 bags

2. sketch of 3 thousands, 7 hundreds, 5 tens

3. a) 3408 **b)** 2310

4. a) e.g., 1 thousand, 4 hundreds, 2 tens, 3 ones
   b) e.g., 14 hundreds, 2 tens, 3 ones
   c) e.g., 14 hundreds, 1 ten, 13 ones

5. 4 thousands + 8 hundreds + 9 ones, or 4000 + 800 + 9

6. 4029

7. a) 2085 **b)** 6256

Lesson 5: pages 48–49
Writing Number Words

1. a) nine thousand nine hundred ninety-five
   b) three thousand nine hundred fifty

2. $3460

3. a) one thousand five hundred
   b) one thousand five

4. a) 6214 **b)** 2100 **c)** 6024

5. e.g., six thousand eight hundred fifty-two is 6852;
eight thousand six hundred fifty-two is 8652;
two thousand eight hundred fifty-six is 2856;
two thousand six hundred fifty-eight is 2658

6. e.g., Banks might want to make sure that someone hasn't added an extra digit to the numeral. They can use the number word to check.

Lesson 6: pages 50–52
Locating Numbers on a Number Line

1. e.g., 1908 just a little bit past 1900, 1952 about halfway between 1900 and 2000, and 1872 closer to 1900 than 1800

2. a) e.g., A: 3875 B: 4000 C: 4050
   b) e.g., A: 3450 B: 4050 C: 4775

3. a)

1200 1210 1220   1260 1270 1280 1290 1300 1310

b)

2000 2500 3000
2045 3045 4045 5045 6045 7045

c)

2000 2100 2200
2350 2450 2550·2650 2750 2850

4. a)–b)

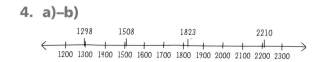

1298   1508   1823   2210
1200 1300 1400 1500 1600 1700 1800 1900 2000 2100 2200 2300

5. a) The number line is missing 1430.
   b) The first number should be 2000, not 2900.
   c) The number line is missing 5500.

6. e.g., To use a number line, you have to know what numbers to put at each mark. The marks are evenly spaced, so it's like counting by a number.

Lesson 7: pages 54–56
Comparing and Ordering Numbers

1. $2473 < 5117$

2. a) $1281 > 654$ b) $6772 < 7276$
   c) $2395 < 2942$ d) $1135 < 1138$

3. e.g., $3225 > 3144$, $3925 > 3874$, $3525 > 3344$

4. a) car bumper stickers   b) pencils

5. a) 7079   b) 9770

6. a) 4346, 4356   b) 7316, 7416, 7516

7. e.g., Look at the thousands digits first. If one is greater, then that whole number is greater. If those digits are the same, then look at the hundreds, and so on.

8. a) e.g., 9350, 1384, 2676
   b) e.g., 9350, 2676, 1384

9. e.g., You start both by looking at the digit with the greatest place value. With 3-digit numbers you start with hundreds, but with 4-digit numbers you start with thousands.

Lesson 8: pages 58–59
Communicating about Ordering Numbers

1. e.g., Some of the numbers are more than 1000 and some are less, so I put 1000 in the middle. I wrote 4-digit numbers before 1000 and the other numbers after 1000: 1876, 1540, 1000, 86, 865. Then I compared the hundreds digits of the numbers more than 1000. The numbers are in the right order because 8 hundreds is greater than 5 hundreds. 86 and 865 are easy to compare because 86 is less than 100 and 865 is greater than 100. So 865 comes before 86. The numbers in order from greatest to least are 1876, 1540, 1000, 865, 86.

2. a) 3869, 3867, 473, 450, 392

Chapter 2 Review pages 60–62

1. a) 30   b) 25   c) 33

2. a) 200   b) 33   c) 330

3. e.g., sketch of 2 thousands, 9 ones

4. 3407

5. a) e.g., sketch of 2 thousands, 2 hundreds, 3 tens
   b) $2000 + 200 + 30$ or 2 thousands + 2 hundreds + 3 tens

6. a) 1096   b) 6129

7. e.g., 1 hundred more than 9900; 1 ten more than 9990

8. a) three thousand one hundred five
   b) eight thousand two

9. e.g.,

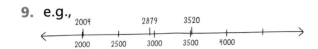

10. a) 2370, 2380, 2400
   b) 4265, 4275

11. a) The number 7000 should be 6100.
   b) The number 3350 should be 3400.

12. e.g., 2295 > 1524,
   8295 > 1504,
   6295 > 1594

13. 6043, 6214, 7053, 7158, 8124

Chapter 3
Addition and Subtraction

Lesson 1: pages 68–69
Solving Problems by Estimating
1. yes; e.g., estimate

2. **a)** about 1400   **b)** about 750

3. **a)** 350; e.g., about 350
   b) 1000; e.g., about 1000

4. no

5. e.g., Often in real life you can use an estimate instead of the exact sum, like when you need to see if you have enough money to buy something.

Lesson 2: pages 70–72
Estimating Sums
1. e.g., about 3300 m

2. **a)** e.g., about 1100
   b) e.g., about 5100
   c) e.g., about 8000
   d) e.g., about 7000

3. **a)** e.g., 2867 is between 2000 and 3000, and 4189 is between 4000 and 5000, so the answer is between 2000 + 4000 = 6000 and 3000 + 5000 = 8000
   b) e.g., estimate 2867 + 4189 as 3000 + 4000 = 7000

4. e.g., about 5000

5. e.g., 2457 + 2503 can be estimated as 2000 + 2000, and your estimate is "more than 4000." You can also add 500 + 500 to get 1000 and add that to 4000, so your estimate is "close to 5000."

Lesson 4: pages 74–76
Adding from Left to Right
1. **a)** yes
   b) e.g., estimated, then calculated

2. **a)** e.g., about 8000; answer is 8208
   b) e.g., about 9000
   c) e.g., about 9000; answer is 8970
   d) e.g., almost 10 000
   e) e.g., about 7000; answer is 7050
   f) e.g., about 9000; answer is 8945

3. **a)** 7051   **b)** yes

4. **a)** e.g., Cole calculated 1000 + 0 and 700 + 700 in two steps, and Julia calculated 1700 + 700 in one step.
   b)

Cole's Calculation	Julia's Calculation
3882	3882
+ 938	+ 938
3000	4700
1700	110
110	+ 10
+ 10	4820
4820	

5. e.g., Add the digits in each place value column: thousands, hundreds, tens, and ones. Grid paper helps keep track of the digits in each column. To add 2568 and 987, add 2000 + 0, 500 + 900, 60 + 80, and 8 + 7.

	2	5	6	8
+		9	8	7
	2	0	0	0
	1	4	0	0
		1	4	0
+			1	5
	3	5	5	5

6. a) 1877 b) 9327 c) 5523 d) 5712

7. e.g., It is similar because you add the digits with the same place value. It is different because you would start with the thousands instead of the hundreds.

Lesson 5: pages 78–80
Adding from Right to Left

1. 3825 ·

2. yes

3. a) e.g., about 3000
   b) e.g., about 5000; sum is 5416
   c) e.g., about 5000; sum is 4810
   d) e.g., about 6500

4. a) e.g., 7642 + 531 = 8173
   b) 1246 + 357 = 1603

5. e.g., Some numbers might be quicker to add from right to left, and some might be quicker to add from left to right.

Mid-Chapter (3) Review pages 81–82

1. a) yes b) e.g., estimate

2. a) 5077 b) 7109

3. a) 4664 b) 1174

4. a) 2006 b) e.g., estimate

5. a) e.g., mental math; 50
   b) e.g., mental math; 5000
   c) e.g., pencil and paper; 9259
   d) e.g., pencil and paper; 7610

6. a) 7524 b) e.g., The sum is between 1000 + 2000 + 3000 = 6000 and 2000 + 3000 + 4000 = 9000. So 7524 is reasonable.

7. a) 6024 b) 6656

Lesson 7: pages 84–86
Subtracting Numbers Close to Hundreds or Thousands

1. 218 years

2. e.g., I am 9 and my grandpa is 65. The difference between my age and my grandpa's is 56.

3. a) 102 b) 702 c) 3002 d) 1008

4. 1538 kg

5. e.g., If you add the same number to both numbers, the difference will be the same. e.g., Add 11 to both numbers.

	3	3	7	1
−	1	7	0	0
	1	6	7	1

6. a) 1693 b) 1002

7. e.g., If you add 1 to 999, you get 1000, which is easy to add on to or subtract from a number. For example, 1632 − 999; add 1 to 999 to get 1000, add on 632 to 1000 to get 1632; 1 + 632 = 633.

Lesson 8: pages 88–90
Regrouping before Subtracting

1. a) 469
   b) e.g., 788 is close to 800. 12 hundreds minus 8 hundreds is 4 hundreds, so 469 seems reasonable.

2. a) 3191   b) 147

3. 2429

4. a) about 1300
   b) about 2100; difference is 2067

5. a) e.g., mental math; 5000
   b) e.g., paper and pencil; 4029
   c) e.g., mental math; 1042
   d) e.g., mental math; 2001

6. 9899

7. a) e.g., Jon started at the left by subtracting 5 hundreds from 20 hundreds, which gave him 15 hundreds or 1500. Then he calculated 1500 − 75 = 1425. The difference is 1425.
   b) 4505

8. e.g., I compare each digit in the top number with the digit in the same place value in the bottom number. If the digit in the top number is less than the digit in the bottom number, I have to regroup. For 2568 − 917, I have to regroup to get more hundreds because I can't subtract 9 hundreds from 5 hundreds. I have enough tens and ones to subtract without regrouping.

Lesson 9: pages 92–93
Subtracting by Renaming

1. 3917 days

2. a) 121 people

3. 2544 points

4. a) e.g., less than 600; 565
   b) e.g., about 1500; 1565
   c) e.g., about 2700; 2722
   d) e.g., just less than 5700; 5668

5. 6086

6. e.g., Method 1: 750 + 250 = 1000, so 1000 − 250 = 750. Method 2: change 1000 to 999 + 1 and subtract.

7. a) He needs to add 1 to 2562. His answer should be 2563.
   b) e.g., 2998 + 2 or 2997 + 3 would allow him to subtract 437 without regrouping. He would have to remember to add 2 or 3 after he subtracted.

Lesson 10: pages 94–95
Communicating about Number Concepts and Procedures

1. 657

2. 738

3. 799

4. e.g., If I explain how I estimated to someone, it will help them understand what I did. If I made an error, that person can show me what I did wrong.

Chapter 3 Review pages 96–98

1. a) 457   b) 896   c) 845

2. a) yes   b) e.g., estimate

3. The walk around Lakeview Park is longer.

4. 7110 books

5. a) e.g., about 4400 m high
   b) 4435 m

6. a) 1707 b) 7874 c) 8116 d) 8343

7. a) 131 b) 225 c) 701 d) 4002

8. 1399 kg

9. a) e.g., about 4000; 4119
   b) e.g., just less than 3000; 2919
   c) e.g., about 1700; 1731
   d) e.g., about 7000; 7252

10. 1721

Cumulative Review:
Chapters 1–3 pages 100–101
   1. D   5. A   9. A   13. D
   2. C   6. B   10. D   14. D
   3. B   7. B   11. A   15. A
   4. B   8. B   12. B

Chapter 4
Data Relationships

Lesson 1: pages 106–109
Interpreting and Comparing Pictographs
   1. a) e.g., They are the same because they both use circles for symbols, have the same titles, and the same categories; they are different because they show different scales.
      b) yes
      c) 60

d) e.g., If you used a scale of 1, you'd have to draw 35 circles for Don't Know. This is too many circles to draw and they probably wouldn't fit on a sheet of paper.

2. a) 90 rattlesnakes b) yes

3. No

4. a) Each symbol means 2 students.
   b) Each symbol means 10 students.

Lesson 2: pages 110–113
Constructing Pictographs
   1. a) e.g.,

   Visitors to the Dinosaur Show
   Mon. ●●●●●●
   Tues. ●●●●●●●●
   Wed. ●●●●●●●●
   Thurs.●●●●●●●●●
   Fri. ●●●●●●●●●●●●●●●●●●

   Each ● means 25 visitors.

   b) e.g., Count by 25 to 225.
   c) e.g., The number of visitors gets greater as the week goes on; twice as many visitors went on Friday as on Thursday.
   d) e.g., There would be a lot more circles and there would have to be some half circles.

2. a) team A; team C
   b)

Team	Number of bones
A	25
B	15
C	10
D	20

   c) e.g., pictograph with scale 1 shape means 10 bones

3. **a)** 1 square means 5 students
   b) 35 students
   c) e.g.,

   Did You Use a Computer Yesterday?

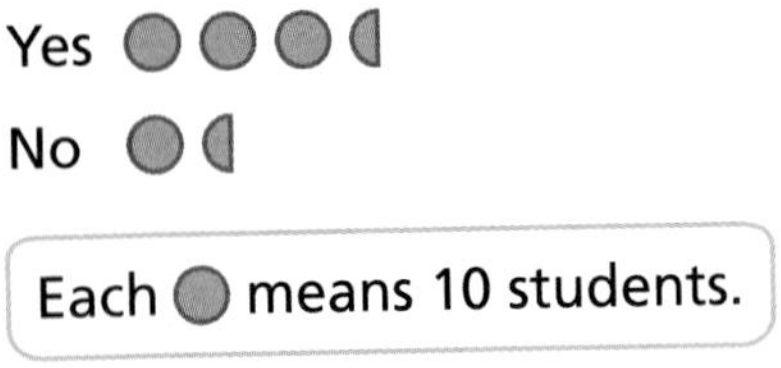

   Yes ● ● ● ◖

   No ● ◖

   | Each ● means 10 students. |

4. **a)** no
   b) no
   c) e.g.,

   Animals for Adoption

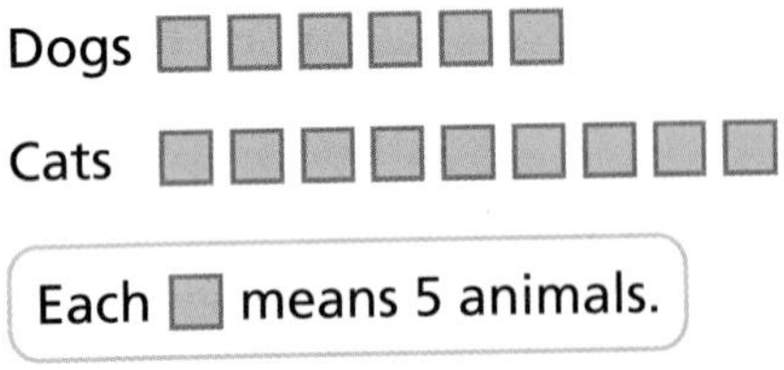

   Dogs ▢ ▢ ▢ ▢ ▢ ▢

   Cats ▢ ▢ ▢ ▢ ▢ ▢ ▢ ▢

   | Each ▢ means 5 animals. |

5. **a)** e.g., the 20 spin experiment
   b) e.g., the 400 spin experiment

6. e.g., I look at the numbers in the data and see if I can count by 2s or 5s or 10s or 20s to those numbers.

Lesson 3: pages 114–117
Interpreting and Comparing Bar Graphs

1. **a)** e.g., same title and labels, and 3 bars each; each shows a different scale
   b) no
   c) e.g., Desmond's graph shows that there are just less than 20 quarters, about 55 $1 coins and about 25 $2 coins. Ken's graph shows that there are just less than 20 quarters, about 50 $1 coins and about 30 $2 coins. Both graphs show that $1 coins were collected most often.
   d) no

2. **a)** Kid Power
   b) Renée's graph has a scale of 5; Amelia's graph has a scale of 8
   c) no
   d) e.g., about 90 students

3. **a)** e.g., about 250 more turbines
   b) e.g., about 650 turbines in Canada
   c) probably
   d) e.g., Keifer's scale is half the size of Ryan's. He needs to double the height for each bar because the height of each square on his graph shows half the number of turbines as the height of each square on Ryan's graph.

4. **a)** e.g., The scale is 50.
   b) e.g., I can use squares and a scale of 1 square for 50 people.
   c) e.g., They both use scales and shapes. But pictographs show rows or columns of shapes while bar graphs show bars made of squares in rows or columns. The squares are connected together while the shapes in a pictograph are not.

Lesson 4: pages 118–120
Constructing Bar Graphs

1. **a)** no
   b)

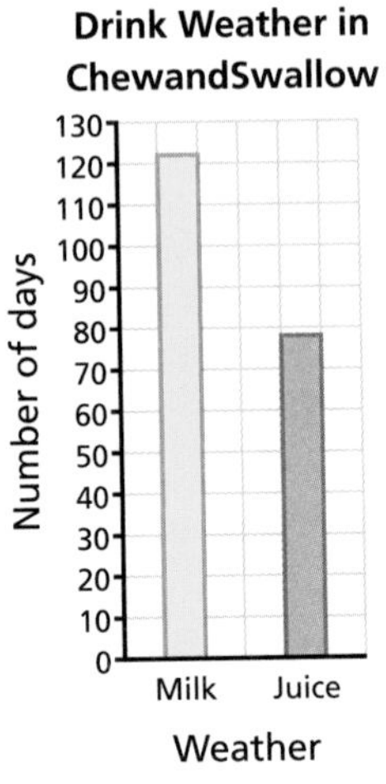

   Drink Weather in ChewandSwallow

   c) e.g., scale of 10

2. a)

Shows at Children's Festival

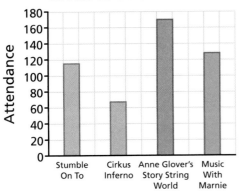

b) e.g., scale of 20

c) The graph; For example, it is easy to compare the bars to see that the bar for Anne Glover's Story String World is about twice as high as the bar for Cirkus Inferno. It's not as easy to see that 170 is about twice as many as 87.

3. a) e.g., scale of 10

b)

Pond Depth

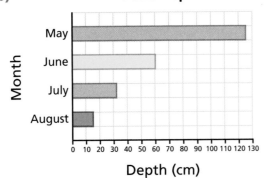

c) e.g., The depth of the pond decreases each month.

Mid-Chapter (4) Review

1. a) e.g., One pictograph has a scale of 1 symbol means 1 can. So only whole cans are needed to show each number of cans.

b) e.g., The other pictograph has a scale of 1 symbol means 10 cans. You need some half cans for 15 and 5.

c) yes

d) e.g., The first graph shows that Tim collected the most cans, followed by Rosie and Aputik. It also shows that Tim collected about the same number of cans as both the other students combined.

2. a)

books	60
newspapers	15
magazines	30
comics	45

b) e.g.,

What Students Are Reading

books	●●●●●●●●●●●●
newspapers	○○●
magazines	●●●●●●
comics	●●●●●●●●

Each ● means 5 students.

c) e.g., I chose a scale of 5 because it's easy to count by 5s, and you need twice as many circles as the graph with the scale of 10. A circle is an easy shape to divide in half if I need to.

d) e.g., 150 students

3. a) e.g., Ethan

b) about 78

c) yes

407

4. a) no

b) e.g.,

Sandwiches Sold

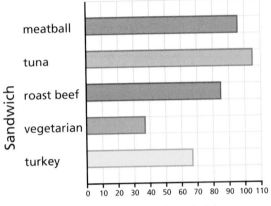

Number Sold

c) e.g., scale of 10

d) e.g., Tuna is the most popular sandwich. Vegetarian is the least popular.

Lesson 6: pages 128–131

Using Venn Diagrams

1. Numbers

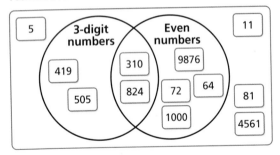

2. a) Numbers 1 to 20

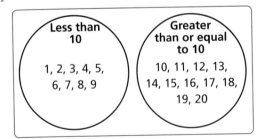

b) no

c) e.g., Even numbers

d) e.g., **Numbers 1 to 20**

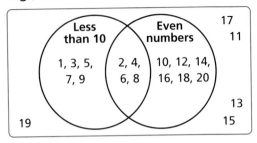

3. a) e.g., "Even numbers" and "10 or greater"

b) e.g., **Numbers 1 to 20**

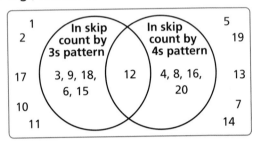

4. e.g., **Number of Students in Each Classroom**

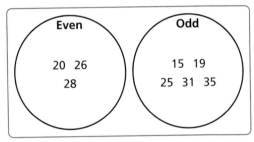

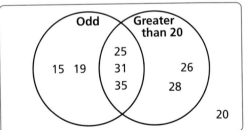

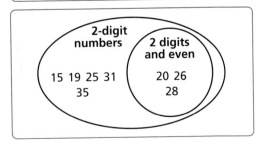

5. a) 2-digit and Even
   b) Even and Odd

Lesson 7: pages 132–133
Using Carroll Diagrams

1. a) e.g.,

	Fewer than 4 digits	4 digits or more
Has the digit 1	317, 108, 118	4871, 1633, 1979
Does not have the digit 1	427, 26, 270	3535, 3988, 3373

   b) no

2. a)–b) e.g.,

	Tens digit is even	Tens digit is not even
3 digits or more	200, 168, 320, 1280, **141**, **5521**	**555, 114**
Fewer than 3 digits	24, 25, **46**, **81**	37, 15, 17, 36, 99, 10, **78**, **38**

3. a) e.g., The labels might be 2 digits for column 1 and 3 digits for column 2. The other labels might be Even and Odd.
   b) e.g.,

	2-digit numbers	3-digit numbers
Even	10, 24, 98, **66, 76**	352, 564, 670, **222, 444**
Odd	19, 91, 99, **13, 15**	103, 255, 787, **435, 437**

4. e.g.,

	2 digits	Not 2 digits
Even	30, 32, 52, 54	106, 4586, 6596
Odd	19, 21	7, 111, 249

Lesson 8: pages 134–135
Solving Problems Using Diagrams

1. 9 female bears are not cubs.

2. 3 triangles are red.

3. e.g., **Shapes**

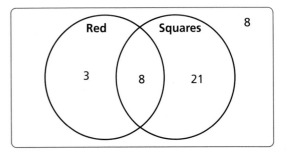

4. e.g., 11 cards are players that are not Canadian and not goalies.

5. a) e.g.,

	Boys	Girls
Wear glasses	9	6
Don't wear glasses	8	5

   b) e.g.,

	Boys	Girls
Wear glasses	9	
Don't wear glasses	8	5

There are 28 students in our class. 9 boys wear glasses. 5 girls don't wear glasses. 8 boys don't wear glasses. How many girls wear glasses?

Chapter 4 Review pages 136–138

1. a) 10 thank-you cards
   b) yes

2. a) e.g., **Favourite Colours**

Red ▪▪▪▪▪▪

Blue ▪▪▪▪▪

Green ▪▪▪

Other ▪▪▪▪

> Each ▪ means 2 students.

b) e.g., scale of 2

3. a) yes
  b) e.g., The number of events is increasing. There were about 25 more events in 2006 than in 1994.

4. a)

Growing Seeds

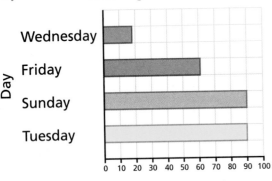

Number Growing

b) e.g., scale of 10
c) e.g., The seeds began to grow slowly and then reached 90 on Sunday. After Sunday, no new seeds seemed to grow.

5. a) Numbers 10 to 25

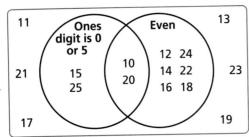

b) yes

6. a)–b) e.g.,

	2-digit numbers	3-digit numbers
Has the digit 5	15, 54, 95, **35, 55**	675, 564, 353, **555, 125**
Does not have the digit 5	30, 19, 61, **36, 10**	103, 249, 787, **111, 222**

7. 14 girls

Chapter 5
2-D Geometry

Lesson 1: pages 144–146
Lines of Symmetry

1. a) C
  b)

A (5 lines), B (6 lines), D (2 lines)

2. a) A and B

3. a) B, D, E, and G

4. B

5. B, C

6. C

Lesson 2: pages 147–150
Using a Symmetry Tool

1. A and C

2. a)–b) e.g.,

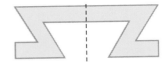

3. A, B, and D

4.

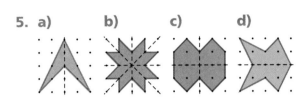

5. a)   b)   c)   d)

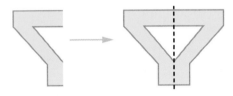

6. a)–b) e.g.,

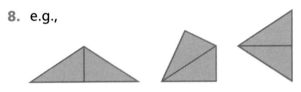

7. a) e.g., all have vertical lines of symmetry

8. e.g.,

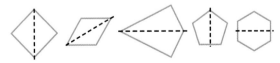

9. e.g., Each side that I trace will be equal to one of the sides on the original shape.

Mid-Chapter (5) Review
pages 152–153

1. e.g., A has 4 lines of symmetry; B has 2; D has 1

2. a) A, B, D, M, O, T, and V

3.

4. a) e.g., chalkboard

Lesson 4: pages 154–156
Counting Lines of Symmetry

1. a) e.g., predictions: J: 1; K: 1; L: 0; M: 0; N: more than 1; O: 1; P: 0
b) number of lines of symmetry: J: 1; K: 1; L: 0; M: 0; N: 3; O: 1; P: 1

2. Number of lines of symmetry: A: 1; B: 2; C: 2; D: 0; E: 6; F: 0; G: 0; H: 0; I: 4; J: 1; K: 8; L: 4

3. a) B   **b)** C

4. e.g., no

Lesson 5: pages 158–160
Communicating about Symmetry

1. e.g., The shape of D is symmetrical and has 2 lines of symmetry.

I tried to find other lines of symmetry by folding but that is all it has. The design has 1 line of symmetry.

2. a) e.g., My first shape has 3 lines of symmetry.
b) e.g., Yes, only shape A has 3 lines of symmetry.

3. a) e.g., The shape has 6 lines of symmetry. Its colour design has 2 lines of symmetry.
b) e.g., My partner knows the shape is C because it's the only shape with 6 lines of symmetry.

c) e.g., I could have provided more details about the colour design. I could have described the colour design as a rhombus located in the centre of the hexagon.

Chapter 5 Review pages 163–164

1. a) e.g., a desk top

2. e.g.,

3. a)–b) e.g., the red pattern block has 1 line of symmetry

4. a) e.g., Trace all the pattern blocks and cut them out. Fold them in all directions to see whether the sides match. If the sides match, then the fold line is a line of symmetry.
b) hexagon has 6; square has 4; triangle has 3; beige rhombus has 2; blue rhombus has 2; trapezoid has 1

5. a) e.g.,

b) e.g.,

c) e.g.,

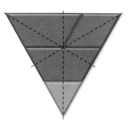

Chapter 6
Multiplication and Division Facts

Lesson 1: pages 170–173
Multiplying by Skip Counting

1. a) Eastside:
$2 + 2 + 2 + 2 + 2 + 2 + 2 = $ ▨ or $7 \times 2 = $ ▨
Gully View:
$2 + 2 + 2 + 2 + 2 + 2 = $ ▨ or $6 \times 2 = $ ▨
Beachside:
$2 + 2 + 2 + 2 + 2 + 2 + 2 + 2 + 2 = $ ▨ or $9 \times 2 = $ ▨
b) Eastside: 14 bikes; Gully View: 12 bikes; Beachside: 18 bikes

2. a) e.g., skip counting by 5s
b) 35 days c) e.g., I could add 7 to the number of days in 5 weeks.

3. 18 wheels

4. a) $6 \times 3 = 18$
b) He can add one more 3 on his number line.

5. 24 legs

6. a) 12 b) 6 c) 48 d) 3 e) 20
f) 6

7. 36 legs

8. a) 9 b) 0 c) e.g., $5 \times 1 = 5$
d) e.g., $5 \times 0 = 0$

9. e.g., $2 \times 6 = 12$; $4 \times 3 = 12$

10. e.g., The order doesn't matter when you multiply 2 numbers. I like to skip count by the easier number. To multiply 7 by 5, I would skip count by 5 because I like to count by 5s. I would continue counting until I counted 7 fives: 5, 10, 15, 20, 25, 30, 35.

Lesson 2: pages 174–175
Building on Multiplication Facts

1. a) e.g., Maybe she knew 5 sixes equal 30 and she could count on from there.
   b) e.g., $5 \times 6 = 30$, $30 + 6 + 6 = 42$, so $7 \times 6 = 42$
   c) e.g., group 7×6 as 2 groups of 7×3

2. a) 24
   b) 32; e.g., 2 more 4s give 8×4.

3. a) 42   b) 45   c) 42   d) 35

4. e.g., I know $8 \times 8 = 64$, and $64 + 8 = 72$ so $9 \times 8 = 72$.

Lesson 3: pages 176–177
Doubling Multiplication Facts

1. a) 12 dancers; 24 dancers
   b) 14 dancers; 28 dancers

2. a) 24   b) 48

3. e.g., Since 4 is double 2, double the number in the row above for each column. For example, $5 \times 2 = 10$, $5 \times 4 = 20$, or double 10. Since 8 is double 4, double the number in the row above for each column. For example, $5 \times 4 = 20$, $5 \times 8 = 40$, or double 20.

4. e.g., Since 6 is double 3, double the number in the row above for each column. For example, $5 \times 3 = 15$, $5 \times 6 = 30$, or double 15.

5. a) $5 + 5 = 10$; $10 + 10 = 20$; $20 + 20 = 40$; so $5 \times 8 = 40$

Lesson 4: pages 178–180
Halving and Doubling Multiplication Facts

1. a) $9 \times 6 = $ ▇
   b)–c)

$4 \times 6 = 24$; $4 \times 6 = 24$; $1 \times 6 = 6$; $24 + 24 + 6 = 54$, so $9 \times 6 = 54$

2. a) 24 batteries
   b) 48 batteries
   c) 56 batteries

3. a) 36   b) 18   c) 30   d) 54   e) 64
   f) 72

4. a) 6   b) 7   c) 4   d) 4

5. a) 16; 40   b) 21; 49   c) 12; 28
   d) 15; 25

6. e.g., $4 \times 3 = 12$; double 12 to get 4×6; double 12 plus 6 to get 4×7

Lesson 6: pages 182–184
Multiplying by 8 and 9

1. 56 students; 63 students

2. 54 squares

3. e.g., Complete the 9× row by subtracting the top number in each column from the number in that column in the 10× row; subtract again to get the number in the 8× row.

4. e.g., $4 \times 9 = 4 \times 10 - 4$; for 4×8, subtract 4 twice

5. a) 40 food tickets   b) 36 food tickets

6. e.g., The 9× facts are close to the 10× facts. You can multiply the number by 10 and then subtract the number.

Mid-Chapter (6) Review
pages 186–187
1. Week 1: 8 cm, Week 2: 10 cm, Week 3: 14 cm, Week 4: 16 cm, Week 5: 18 cm

2. e.g., Skip count by 5s eight times to get 40.

3. a) 7   b) skip 2 more 5s

4. a) $24   b) $48

5. a) 16; 32   b) 18; 45   c) 24; 48
   d) 27; 63

6. a) 40   b) 56   c) 14   d) 27

7. a) 0   b) 1   c) 0   d) 0

Lesson 7: pages 188–190
Sharing and Grouping
1. a) Problem 1: 6 kg
   b) Problem 2: 7 weeks

2. 4

3. 5

4. a) 3   b) 7   c) 6   d) 7

5. a) 4   b) 9

6. a) e.g., The number line shows that 7 jumps of 4 equal 28, which is the same as dividing 28 into groups of 4 or calculating $28 \div 4$.
   b) e.g., Use a number line to show jumps of 5 from 0 to 35, so that there are 7 jumps.

7. e.g., There are 56 flowers. Sammie wants to plant an equal number of flowers in 8 rows. How many flowers will be in each row? (Answer: 7)

8. e.g., Use skip counting forward or backwards to calculate $35 \div 5 = \blacksquare$.
   Forward: Count the number of 5s to equal 35.
   5, 10, 15, 20, 35, 30, 35
   Count 7 fives so $35 \div 5 = 7$.

Lesson 8: pages 192–195
Division and Multiplication
1. 9 circles

2. $4 \times 8 = 32$, $32 \div 8 = 4$

3. a) 9   b) 9   c) 8   d) 8   e) 9   f) 7

4. 6 adults

5. 8 beaters

6. 6 packs

7. a) $45 \div 9 = 5$
   b) $9 \times \blacksquare = 45$; $\blacksquare = 5$

8. a) e.g., 7, 4, 3; $7 \times 1 = 7$; $7 \div 1 = 7$

9. a) All the missing numbers are 0.
   b) e.g., The answer is always 0.

10. a) $3 \times 9 = 27$
    b) $6 \times 8 = 48$; The answer should be 9.
    c) $8 \times 8 = 64$
    d) $9 \times 8 = 72$; The answer should be 9.

11. a) 5 b) 25 c) 9 d) 63 e) 9 f) 7

12. 30 ÷ 5 = 6; 30 ÷ 6 = 5

Lesson 10: pages 198–199
Solving Problems by Working Backwards

1. a)
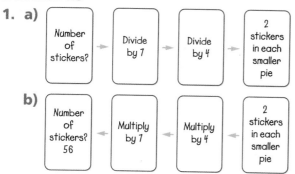

   b)

2. 36 cm

3. $15

4. e.g., Suzy spent $28 dollars at the book store. She bought some books for $6 each. Then she bought 2 bookmarks for $2 each. How many books did she buy? (Answer: 4 books)

Chapter 6 Review pages 200–202

1. a) 8 × 5 = 40
   b) e.g., Show 5 jumps of 8 to get 40.

2. a) 30 b) 42 c) 36

3. a) 24 pens b) 28 pens

4. a) 3 × 4 = 12 beads b) 24 beads

5. a) 21; 42 b) 24; 54

6. a) 20 cents b) 45 cents

7. a) 70 players
   b) 63 players

8. 7 groups

9. 8 books

10. a) 9 × 8 = 72 b) 6 × 8 = 48
    c) 7 × 4 = 28 d) 9 × 6 = 54

11. a) 7 b) 0 c) 6 d) 1

12. 8 people

Chapter 7
Fractions and Decimals

Lesson 1: pages 208–211
Fractions of a Whole

1. a) e.g.,
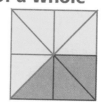

   b) $\frac{5}{8}$

2. a) triangle b) e.g., $\frac{3}{5}, \frac{5}{6}, \frac{5}{8}$

3. a) $\frac{3}{5}$ b) $\frac{2}{5}$

   e.g.,

4. a) $\frac{1}{3}$ b) $\frac{2}{6}$

5. a) e.g.,

   b) e.g., $\frac{4}{12}$

6. a)

   b) $\frac{3}{10}$

7. e.g., blue, green, and yellow

8. e.g., The apples are different sizes.

9. a) 4 of 10 parts are yellow. b) 7 of 10 parts are yellow and green.
   c) 10 of 10 parts are coloured.
   d) 0 of 10 parts are purple.

10. e.g., blue shows $\frac{2}{10}$;
    blue and yellow show $\frac{6}{10}$;
    blue, yellow, and green show $\frac{9}{10}$

11. a) $\frac{1}{3}$ green, $\frac{2}{3}$ red
    b) $\frac{4}{9}$ clouds, $\frac{5}{9}$ clear sky

12. e.g., The cupboard doors are $\frac{2}{3}$ blue and $\frac{1}{3}$ white.

Lesson 2: pages 212–214
Fractions of a Group

1. a) e.g., $\frac{2}{4}$ live on land, $\frac{1}{4}$ have tusks, $\frac{3}{4}$ have no tusks
   b) $\frac{1}{4}, \frac{2}{4}, \frac{3}{4}$

2. a) $\frac{3}{4}$ b) $\frac{2}{5}$ are running

3. a) e.g., 1 pencil and 2 pens, circle the 2 pens to make $\frac{2}{3}$
   b) e.g., 3 pencil crayons

4. b) $\frac{1}{6}, \frac{2}{6}, \frac{3}{6}$

5. e.g., The one with the greater numerator is greater.

6. e.g., I don't know unless I know the number of strawberries they each started with.

7. e.g., They both show $\frac{1}{6}$ and $\frac{5}{6}$. One model is one whole shape divided into 6 equal parts but the other is a whole group made up of different shapes and sizes.

Lesson 4: pages 216–218
Comparing and Ordering Fractions

1. a) Cory b) Aneela c) Lang

2. $\frac{1}{8}, \frac{1}{5}, \frac{1}{3}, \frac{1}{2}$

3. a) $\frac{1}{2}$ b) $\frac{1}{3}$ c) $\frac{1}{2}$

4. $\frac{3}{4}$

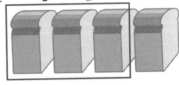

$\frac{3}{8}$

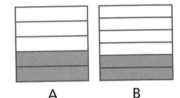

5. $\frac{4}{10}, \frac{4}{8}, \frac{4}{5}, \frac{4}{4}$

6. true

7. a) and b)

```
┌─────────┐   ┌─────────┐
│         │   │         │
│         │   │         │
│         │   │         │
│▓▓▓▓▓▓▓▓▓│   │▓▓▓▓▓▓▓▓▓│
│         │   │         │
└─────────┘   └─────────┘
     A             B
```

c) $\frac{2}{5}$ is greater d) less

8. a) $\frac{2}{8}$ is after b) $\frac{5}{8}$ is after

9. $\frac{5}{10}, \frac{5}{8}, \frac{5}{12}$

10. e.g., B

11. e.g., The sizes and shapes of the fractions are different, so it's hard to compare.

Lesson 5: pages 220–222
Using Benchmarks to Order Fractions

1. a) $\frac{6}{10}, \frac{7}{8}$
   b) $\frac{1}{5}, \frac{4}{10}, \frac{6}{10}, \frac{7}{8}$

2. a) $\frac{9}{10}$ **c)** $\frac{1}{8}, \frac{1}{5}, \frac{3}{8}, \frac{7}{12}, \frac{9}{10}$

3. a) e.g., $\frac{1}{3}, \frac{1}{4}, \frac{1}{5}$ **b)** e.g., $\frac{2}{3}, \frac{3}{4}, \frac{5}{6}$

4.

$\frac{2}{10}$

0     $\frac{1}{2}$ 1

5. a) e.g., $\frac{2}{5}$ is less than $\frac{1}{2}$.

b) e.g., $\frac{3}{5}$ is between $\frac{1}{2}$ and 1.

Lesson 6: pages 224–226
Solving Problems by Drawing Diagrams

1. e.g., There could be 0, 1, or 2 dark-haired girls wearing jackets.

2. e.g.,

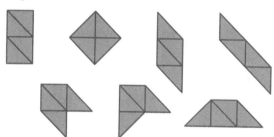

3. $\frac{3}{10}$

4. 2 hours

5. 0 to 6 times

6. e.g., $\frac{2}{6}$ of a group of puppies have brown spots, $\frac{5}{6}$ have collars on, and $\frac{3}{6}$ are female. How many of the puppies could be females with collars and brown spots?

Mid-Chapter (7) Review
pages 228–229

1. a) $\frac{2}{5}, \frac{3}{5}$ **b)** $\frac{8}{10}, \frac{2}{10}$

2. e.g.,

a)

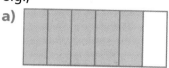

b)

3. a) e.g., $\frac{4}{5}$ are black; $\frac{2}{5}$ have bows; $\frac{1}{5}$ is white

b) $\frac{1}{5}, \frac{2}{5}, \frac{4}{5}$

c)

4. a) $\frac{3}{5}$

b) $\frac{6}{10}$

c) $\frac{3}{4}$

d) $\frac{1}{3}$

5. $\frac{1}{10}, \frac{1}{6}, \frac{1}{5}, \frac{1}{3}$

6. a) closest to $\frac{1}{2}$ **b)** closest to 0
c) closest to 1 **d)** closest to 0

Lesson 7: pages 230–232
Decimal Tenths

1. e.g., $\frac{2}{10}$ or 0.2 of the rings have hearts, $\frac{8}{10}$ or 0.8 are not blue, $\frac{1}{10}$ or 0.1 has a spider, etc.

2. a) $\frac{3}{10}$, 0.3 **b)** $\frac{10}{10}$, 1.0 **c)** $\frac{1}{10}$, 0.1

d) $\frac{0}{10}$, 0.0

3. a) 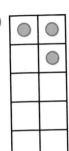 b)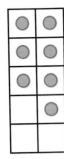

4. 1.0

5. yellow in A, green in B, red in D

6. a)–b) e.g., 0.2 plain; 0.5 with stones; 0.1 red

7. e.g., The pictures for 0.8 and 0.2 both have 10 parts. For 0.8, 8 of the 10 parts are coloured, but for 0.2, only 2 parts are coloured.

Lesson 8: pages 233–235
Decimal Hundredths

1. a) $\frac{69}{100}$, 0.69 b) $\frac{6}{100}$, 0.06

2. a) $\frac{36}{100}$, 0.36 b) $\frac{58}{100}$, 0.58
   c) $\frac{25}{100}$, 0.25 d) $\frac{100}{100}$, 1.00

3. a) 0.89 b) 0.09 c) 0.30 or 0.3
   d) 0.12

4. a) $\frac{67}{100}$ b) $\frac{29}{100}$ c) $\frac{40}{100}$ d) $\frac{4}{100}$

5. a) 31 cm
   b) 1.0 and 1.00

6. a) true b) false; 0.18 c) false; 0.01
   d) true

7. e.g., 0.20 of the numbers end in a 0 or 5; 0.09 of the numbers have double digits (i.e., 33, 55)

8. a)–b) e.g., $\frac{30}{100}$ or 0.30 yellow; $\frac{14}{100}$ or 0.14 blue; $\frac{35}{100}$ or 0.35 green; $\frac{12}{100}$ or 0.12 red

c) green

9. e.g., Out of 100 people, 42 have brown hair.

Lesson 9: pages 236–238
Representing Decimals with Coins

1. a) Julia: 45 pennies, Luis: 4 dimes, 5 pennies
   b) Julia: 92 pennies, Luis: 9 dimes, 2 pennies
   c) Julia: 208 pennies, Luis: 2 loonies, 8 pennies
   d) Julia 120 pennies, Luis: 1 loonie, 2 dimes

2. a) $0.34 b) $2.61

3. a) 2 dimes and 4 pennies, or 24 pennies b) 3 pennies
   c) 9 dimes and 2 pennies, or 92 pennies
   d) 4 dimes or 40 pennies

4. a) $0.32 b) $0.01 c) $0.40
   d) $1.90 e) $2.11 f) $3.24

5. 4 dollars, 40 cents, 4 cents

6. a) $0.90 b) $0.09

7. e.g., There are 100 cents in a dollar, so 1 cent is $\frac{1}{100}$ of a dollar.

Lesson 10: pages 240–242
Estimating Decimal Sums and Differences

1. e.g., about $25

2. e.g., about $2

3. a) about $4 b) about $3

4. no

5. a) e.g., Fruit salad is about $1 more than milk.
   b) e.g., about $3

6. e.g., Kim bought soup and juice. Dalton bought a pita and milk. About how much more money did Dalton spend than Kim? Solution: Kim spent about $1.40 + $1.20 = $2.60. Dalton spent about $2.75 + $1.00 = $3.75. $3.75 − $2.60 = $1.15, so Dalton spent about $1.00 more than Kim did.

7. e.g., agree

8. e.g., $12 + $8 + $1 or $13 + $8

Lesson 11: pages 243–245
Using Mental Math

1. a) e.g., camera and backpack; $12.00 + $20.00 + $1.50 − 1¢ = $33.49
   b) $50.00 − $30.00 − $3.00 − 50¢ + 1¢ = $16.51

2. a) e.g., When the cents are the same, you just need to subtract the dollars. $14.11 − $9.11 = $5.00
   b) e.g., She knows that if she spent $9.11, then she'd have $5.00 left. But she actually spent 4¢ more ($9.15), so she only has $4.96 left.

3. $11.44

4. e.g., I had $25.00 and I spent $22.75. $25.00 − $22.00 = $3.00. I spent 75¢ more, so I have $2.25 left.

5. a) $17.98   b) $10.20   c) $44.00
   d) $7.50

6. a) $20.94   b) $4.06

7. e.g., A

Lesson 12: pages 246–247
Making Change

1. a) e.g., about $20.00   b) $18.01

2. a) about $4; $3.73
   b) about $3; $3.02

3. e.g., I would use the $20 bill to buy the inline skates. My change would be 2¢.

4. e.g., Count up from the cost of the item to the amount you use to pay for it.

Lesson 13: pages 248–250
Adding and Subtracting Decimals

1. a) e.g., about 1.1 m   b) 1.11 m

2. e.g., about 1.3 m; 1.25 m

3. 2.9 m

4. a) $4.01   b) $1.36

5. a) 7.5 kg   b) 12.46 kg

6. e.g., disagree

7. e.g., Adding and subtracting decimals is just like adding and subtracting whole numbers because adding is like counting up and subtracting is like counting down. The only difference is that you count up and down by 10ths or 100ths instead of by whole numbers.

Chapter 7 Review pages 251–254

1. a) $\frac{3}{10}$ b) $\frac{3}{8}$ c) $\frac{3}{5}$

2. A

3. e.g., $\frac{2}{5}$ of the shapes are red; $\frac{0}{5}$ of the shapes are circles; $\frac{3}{5}$ of the shapes are rectangles

4. e.g.,

⬤⬤◯◯
⬤⬤⬤◯

5. $\frac{2}{12}, \frac{2}{6}, \frac{2}{5}, \frac{2}{3}$

6.

$\frac{1}{12}$ $\frac{3}{10}$ $\frac{4}{10}$ $\frac{10}{12}$

0                    $\frac{1}{2}$ 1

7. 1 or 2

8. a) 0.7, $\frac{7}{10}$ **b)** 0.32, $\frac{32}{100}$

9. a) $0.19   **b)** $0.50

10. a) e.g., about $11   **b)** e.g., about 5

11. a) $9.98   **b)** $15.51

12. $1.25

13. 2.27 m

Cumulative Review: Chapters 4–7

pages 256–257

1. D	**5.** C	**9.** A	**13.** C
2. C	**6.** B	**10.** C	
3. A	**7.** C	**11.** D	
4. B	**8.** A	**12.** A	

Chapter 8
Measurement

Lesson 1: pages 262–264
Telling Time to the Hour

1. five o'clock in the afternoon, 5:00 p.m.

2. a) eleven o'clock in the morning, 11:00 a.m.
   b) six o'clock in the morning, 6:00 a.m.

3. a) 4:00 p.m.   **b)** 8:00 p.m.

4. 7 hours

5. a) e.g., starting school
   b) e.g., finishing homework

6. e.g., The times have the same value, but one of them is in the morning and the other is in the afternoon.

7. 10 hours

8. e.g., If the clock was showing 7:00, we would not know if it was in the morning or the evening.

9. a) e.g., with hands   **b)** with digits

Lesson 2: pages 266–267
Time to the Half Hour and Quarter Hour

1. a) 4:15, a quarter after four
   b) 8:45, a quarter to nine
   c) 1:30, half past one

2. a) 12:30, half past twelve
   b) 2:15, a quarter after two
   c) 7:45, a quarter to eight

3. 15 minutes

4. a) e.g., It's a quarter to an hour.
   b) It's nine o'clock.

Lesson 3: pages 268–270
Telling Time to 5 Minutes

1. 10:25

2. a) 12:05; five after twelve
   b) 9:40; twenty to ten
   c) 8:20; twenty after eight

3. a)   **b)**

c)

4. 6:55

5. 25 minutes

6. a bit after 4:30

7. e.g., Ten to three is a useful thing to say because the time is getting close to 3:00 and someone might want to know how many minutes are left. Fifty to three isn't as useful because it's still a long time until 3:00.

8. e.g., The marks on a clock show every 5 minutes, so you can count by 5s to figure out the number of minutes before or after an hour.

Lesson 4: pages 272–275
Telling Time to 1 Minute

1. between the 1 and the 2 but closer to the 2

2. **a)** 8:33
   b) 33 minutes after 8; 27 minutes to 9

3. **a)** 9:08 **b)** 1:54 **c)** 2:16

4. **a)** eighteen minutes to four
   b) one minute to six
   c) twenty-six minutes to seven

5. 1 hour and 50 minutes

6. e.g., Nicky's watch because you know that 9:30 is when the minute hand reaches the 6.

7. e.g., Jon's watch because the concert started at 8, and it's 57 minutes past 8.

8. 3:00

9. **a)–b)** e.g., 4:26 a.m., 6:24 a.m., 2:46 p.m., 6:42 p.m.
   c) e.g., At 4:26 a.m., I'll be sleeping. At 6:42 p.m., I'll take my dog for a walk.

10. e.g., Lunchtime lasts from 11:50 until 1:05. How long is that?

11. e.g., sketch of clock showing 2:16: First look at the hour hand. It's between 2 and 3, so I know the time is between 2:00 and 3:00. Then I look at the minute hand. It's pointing just past the 3. I know it would be 2:15 if it was right on the 3, so I need to count the minute marks to see how many minutes it is after 2:15. It's 1 mark, so the time is 2:16.

Lesson 5: pages 276–278
Writing Dates and Time

1. **a)** September 19, 1999 **b)** 7:19 p.m.

2. **a)** 1959-01-12 **b)** 1989-03-15

3. **a)** July 1, 1867 **b)** October 30, 2004

4. 2008-03-15; 03/15/2008; 15-03-2008

5. **a)** 10:00 p.m. **b)** 6:15 a.m.

6. **a)** e.g., 07:00 **b)** e.g., 15:40

7. e.g., 24-hour time just continues when it reaches 12 noon, so on a timeline you can continue to count up from 12 noon.

8. **a)** no **b)** yes **c)** no **d)** yes

9. **a)** March 17, 2008
   b) e.g., You know there are not 17 months.
   c) e.g., write the day/month/year

d) It could mean March 6, 2004, June 3, 2004, or March 4, 2006.
e) e.g., Sometimes the date and month can be confused.

10. e.g., The advantage with 24-hour times is that you won't mix up a.m. and p.m. if you forget to write a.m. or p.m. The disadvantage of the 24-hour times is that it's hard to figure out what an afternoon time like 14:00 means because our watches and clocks usually only go up to 12.
An advantage with numeric dates is that when the dates are all numbers, it's quick to write them and it's easy to compare them. A disadvantage is that because people do them different ways, you can get mixed up as to what each number means.

Mid-Chapter (8) Review
pages 280–281

1. e.g., At 2:00 p.m. tomorrow, I'll be in school.

2. **a)** 3:00, 3 o'clock
   b) 4:30, half past four
   c) 2:15, a quarter after two
   d) 6:50, ten minutes to seven
   e) 7:36, twenty-four minutes to eight
   f) 6:12, twelve minutes after six

3. **a)** June 4, 1887   **b)** 1907-04-19

4. **a)** 20:25   **b)** 11:48

5. **a)** 5:36 p.m.   **b)** 2:20 a.m.

Lesson 6: pages 282–284
Measuring with Area Units

1. **a)–b)** e.g., 12 square pattern blocks

2. **a)** e.g., notebook and my textbook; The area of my notebook was about 80 square units.
   b) e.g., about 100 square units
   c) e.g., textbook, about 20 square units

3. **a)** e.g., I chose the square and hexagon pattern blocks. Using the square, the area of the cover is about 110 pattern block units.
   b) e.g., Since the second unit is about 2 times as big, I'll use about half as many units to cover the area. The actual measurement was about 60 hexagon block units.

4. e.g., no

5. e.g., The unit tells how big the measurement is.

6. e.g., Squares fit side by side with no gaps and cover the whole area.

Lesson 7: pages 286–288
Counting Square Units

1. **a)** e.g., white   **b)** yellow: 31 square units; green: 28 square units
   c) 40 square units
   d) green, yellow, white

2. **a)** C is 9 square units, A is 12 square units, T is 7 square units   **b)** T, C, A
   c) red is 15 square units, yellow is 13 square units, white is 63 square units
   d) yellow, red, white

3. e.g.,

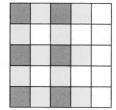 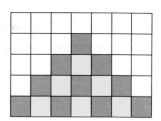

4. e.g., The blue area is 10 square units, the green area is 9 square units, and the white area is 23 square units.

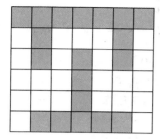

5. e.g., M: 11 square units, P: 9 square units

Lesson 8: pages 290–293
Using Square Centimetres

1. a) e.g., B
   b) e.g., A: about 16 cm²; B: about 18 cm²
   c) A: about 14 cm²; B: about 16 cm²

2. a) e.g., about 25 cm²; about 22 cm²
   b) e.g., about 20 cm²; about 20 cm²
   c) e.g., about 12 cm²; about 9 cm²
   d) e.g., about 20 cm²; about 20 cm²

3. a) A b) e.g., A: about 20 cm², B: about 11 cm²

4. a) 100 cm²
   b) e.g., Place the block on the paper and estimate how many times it would go side by side to cover the paper.
   c) a CD (about 100 cm²)

5. a)–b) e.g., The area of a ruler is about 90 cm²; measured to the nearest whole centimetre, the area is 93 cm². The area of a book cover is about 200 cm²; measured to the nearest whole centimetre, the area is 247 cm².

6. a)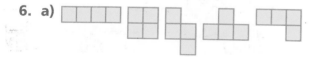

b) 20 cm²
c) 20 cm²

7. e.g., 32 cm²

8. e.g., a square 5 cm by 5 cm

9. no

Lesson 9: pages 294–295
Using Square Metres

1. e.g., board; 8 m²

2. a) e.g., square metres
   b) e.g., square centimetres
   c) e.g., square metres
   d) e.g., square centimetres

3. e.g., a beach towel

4. e.g., a bath mat, my desktop

5. e.g., 1 m²

Lesson 11: pages 298–299
Solving Problems Using Organized Lists

1. 5: 1 by 36, 2 by 18, 3 by 12, 4 by 9, 6 by 6

2. 4: 1 by 40, 2 by 20, 4 by 10, 5 by 8

3. e.g., Carmen could have 50¢, 35¢, 30¢, 26¢, 20¢, 15¢, 11¢, 10¢, 6¢, or 2¢.

4. e.g., How many different rectangles that are 3 tiles wide can you make using up to 36 tiles? (Answer: 12 rectangles)

Lesson 12: pages 300–301
Estimating Areas on Grids

1. e.g., about 240 cm²

2. e.g., computer mouse; about 70 cm²

3. e.g., math book cover about 420 cm²

4. no

Chapter 8 Review pages 302–304

1. **a)** 6:00 a.m. **b)** 10:00 p.m.

2. **a)** 8:00, eight o'clock
   b) 5:30, half past five
   c) 3:45, a quarter to four
   d) 1:25, twenty-five after one
   e) 9:12, twelve minutes after nine
   f) 6:59, one minute to seven

3. 1961-01-26; 07:45

4. e.g., 2008-11-11, 11:00

5. e.g., square tiles

6. **a)** e.g., T **b)** 79 square units

7. **a)** e.g., the area of the cover of my journal, my pencil case, and a sheet of paper
   b)–c) e.g., pencil case: 180 cm^2; journal cover: 336 cm^2; paper: 602 cm^2

8. **a)** e.g., square centimetres
   b) e.g., square metres

9. **a)** 1 by 32, 2 by 16, 4 by 8

10. e.g., I cut out an oval about 15 cm^2.

Chapter 9
Multiplying Multi-Digit Numbers

Lesson 2: pages 312–313
Multiplying 10s and 100s

1. 700 beads, 350 pins

2. **a)** 70 **b)** 120 **c)** 200 **d)** 1800

3. 30 trips

4. **a)** 4 **b)** 6 **c)** 8 **d)** 100

5. $120

6. e.g., 6 × 30 is 6 x 3 tens, so that's 18 tens or 180.
   6 × 3 hundreds is 18 hundreds or 1800.

Lesson 3: pages 314–317
Multiplying Using Arrays

1. **a)** $5 \times 14 = 5 \times 6 + 5 \times 8$ **b)** 70

2. **a)** e.g., sketch of a 7-by-5 array and a 7-by-9 array
   b) 98 squares

3. **a)** 2; 60 + 12; 72
   b) 6; 7×5; 42 + 35; 77
   c) 5×20; 5×7; 100 + 35; 135

4. 126 trees

5. **a)** 126 **b)** 8; 64 **c)** e.g., 30; 2; 160
   d) e.g., $5 \times 20 + 5 \times 8$; 140

6. **a)** 105 carrots **b)** 108 trading cards
   c) 48 pencil crayons **d)** 153 new cars
   e) 128 bottles **f)** 57 stamps

7. **a)** 51 **b)** 32 **c)** 60 **d)** 112 **e)** 126
   f) 207

8. more flowers

9. **a)** a 5-by-20 array and a 5-by-3 array
   b) two 5-by-10 arrays and a 5-by-3 array **c)** three 5-by-7 arrays and a 5-by-2 array

10. e.g., sketch of 4-by-20 array with a 4-by-19 array coloured

11. A, B, D

Lesson 4: pages 318–321
Multiplying Using Expanded Form

1. e.g., Show 5 groups of 6 tens and 7 ones; that's 30 tens and 35 ones. When I regroup, that's 3 hundreds, 3 tens, and 5 ones, which is 335.

2. 128 pieces

3. a) 496 b) 444

4. 425 pencils

5. a) e.g., 6×6 tens is 36 tens, which is more than 350 b) 384 beads

6. a) 85 b) 133 c) 336 d) 200

7. 84 books

8. a) 292 b) 144 c) 204 d) 184

9. FIX

10. a) 145 b) 48 c) 168 d) 261

11. 43

12. 45; 63; 72

13. 2 or 3

14. e.g., adding 25 to 8 is putting together 1 group of 25 and 1 group of 8; multiplying 25 by 8 is putting together 25 groups of 8

Lesson 5: pages 322–324
Estimating Products

1. e.g., yes

2. a) e.g., $9 \times 50 = 450$
   b) e.g., $4 \times 350 = 1400$

3. a) e.g., $6 \times 130 = 780$
   b) e.g., $7 \times 100 = 700$

4. A and D

5. about 300 dogs

6. a) e.g., calculate; yes
   b) e.g., estimate; yes
   c) e.g., estimate; no

7. a) no b) no c) no d) no

8. e.g., An exact answer is not always needed. For example, if you want to know if you have enough money to buy some things or how many people are going to be at an event.

9. e.g., to check if the answer is reasonable

Mid-Chapter (9) Review
pages 326–327

1. A

2. a) 60 b) 100 c) 300 d) 5600

3. 96 pieces

4. a) 54 b) 66 c) 284 d) 85

5. 133 chopsticks

6. 81 buttons

7. e.g., She knew $8 \times 70 = 560$, so 8×72 must be more.

Lesson 6: pages 328–329
Communicating about Solving Problems

1. a) e.g., Carolyn explained that she understood the problem by making a table of the information.
   b) e.g., How did you know you needed to calculate 19×4? How did you calculate 19×4? What does 76 stand for?

2. e.g., 91 human years

Lesson 7: pages 330–332
Multiplying 3-Digit Numbers

1. a)
$$\begin{array}{r} 300 + 20 + 7 \\ \times\ 5 \\ \hline 1500 \\ 100 \\ +\ 35 \\ \hline 1635 \end{array}$$
b)
$$\begin{array}{r} 327 \\ \times\ 5 \\ \hline 35 \\ 100 \\ +1500 \\ \hline 1635 \end{array}$$

2. 876 books

3. a) 1981   b) 4221

4. 1050 carrots

5. a) 2527   b) 1684   c) 1854   d) 1998

6. 1455 balloons and 250 flags

7. a) about 3000; 2958
   b) about 1000; 905
   c) about 2100; 2324
   d) about 4500; 4425

8. 2380 g

9. a) 1000   b) 3765   c) 4326   d) 825

10. e.g., Calvin is trying to raise money for a charity, and has asked every student in his grade to bring in $8. If there are 217 students in Calvin's grade, how much money will he raise? (Answer: $1736)

11. e.g., It is the same because you multiply the number in each place value by 5 and add the products. It is different because you multiply hundreds, tens, and ones instead of just tens and ones.

Lesson 8: pages 334–337
Multiplying Another Way

1. a) 434   b) 1015

2. estimate: 450; answer: 435

3. a) estimate 1500; 1473
   b) estimate 5700
   c) estimate 5400
   d) estimate 9000

4. 2640 cm

5. $4 \times 333 = 1332$

6. a) high
   b) e.g., 4×412; 4×375; 4×420

7. a) 1220   b) 1300   c) 1758   d) 1716

8. 524 passengers

9. a) 2025   b) 2589   c) 4158   d) 7544

10. 5136

11. a) about 6 years old
    b) about 1113 days

12. e.g., Olivia is training for a dance tournament and tries to do 130 sit-ups in a day, 5 days a week. How many sit-ups does Olivia try to do in one week? (Answer: 650)

13. e.g., You can model them both with base ten blocks in the same way. The product is the same. They are different because the regrouping is recorded differently.

Lesson 9: pages 338–340
Choosing a Method to Multiply

1. a) e.g., calculate   b) e.g., estimate

2. a) e.g., multiply 6×33 b) e.g., 9×44

3. a) e.g., mental math
   b) e.g., use base ten blocks
   c) e.g., use mental math

4. a) e.g., calculate   b) e.g., estimate
   c) e.g., calculate

5. e.g., A and B

6. C and D

7. a) e.g., calculate   b) e.g., estimate
   c) e.g., estimate

8. e.g., 10 weeks , 11 weeks, or 20 weeks

Chapter 9 Review pages 342–344

1. a) 6   b) 8   c) 6

2. a) 6×17
   b) e.g., 6-by-10 array and 6-by-7 array
   c) $6 \times 17 = 6 \times 10 + 6 \times 7$

3. 140 items

4. a) 80×9 b) 50×7

5. a) e.g., calculate   b) e.g., estimate

6. 2555 days

7. a) 147   b) 616   c) 880   d) 2454

8. 2408 m

9. 608 m

10. e.g., $750 = 10 \times 75$; $750 = 3 \times 250$;
    $750 = 5 \times 150$

11. a) 315   b) 1962   c) 740   d) 2436

12. a) 3; 4   b) 3   c) 5; 20   d) 2; 2

13. e.g., Heather knows that 100×2 is 200, and 2×4 is 8, so it's easy to see that 2×104 is 208.

14. a) 200; e.g., mental math
    b) 1872; e.g., pencil and paper
    c) 3832; e.g., pencil and paper
    d) 2688; e.g., pencil and paper

Chapter 10
Dividing Multi-Digit Numbers

Lesson 3: pages 352–355
Using Subtraction to Divide

1. a) 16 bunches   b) 3 balloons

2. a) 12 bunches   b) 2 balloons

3. a) 14   b) 10 R2   c) 12 R2   d) 48

4. a) 13 pairs   b) 21 pairs   c) 46 pairs

5. a) yes   b) e.g., Anna's

6. 17 rows

7. a) e.g., Kate had 75 balloons. How many bunches of 8 could she make?
   b) $75 \div 8 = 9$ R3. She could make 9 bunches with 3 balloons left over.

8. e.g., Instead of subtracting 1 group at a time, which will take a long time, subtract 10 groups at a time.

9. 27 packages

10. e.g., $30 \div 4 = 7$ R2 and $37 \div 5 = 7$ R2

11. e.g., If the number can be divided by 4, you can make 4 equal groups. The number is even. If you have 3 left over, the number must be odd.

12. e.g., 8, 14

13. a) e.g., Tien jumped 11 jumps of 3 to 33 and 1 more to get to 34
    b) e.g., Joshua subtracted 3 eleven times, but there was still 1 left.

c) e.g., Cole knew that 30 was 10 threes, so he took 30 away from 34. He still could take away 1 more three (to make 11 threes) and there was still 1 left.

14. e.g., I keep taking 3s away from 59 and see how many 3s to take away and whether there is a remainder. First, I would take away 10 threes and that would leave 29 to make into threes. Then I would take away 5 threes and that would leave 14 to make into threes. Next I use 4 groups of 3. Altogether, I made 19 groups of 3, and there is a remainder of 2.

Lesson 4: pages 356–357
Dividing by Renaming

1. **a)** e.g., divide 92 by 4
   b) e.g., $92 = 80 + 12$
   c) 23 cards

2. **a)** e.g., $67 = 60 + 7$
   b) e.g., $81 = 50 + 30 + 1$
   c) e.g., $57 = 50 + 16 + 1$
   d) e.g., $75 = 60 + 12 + 3$

3. **a)** 15   **b)** 14 R2   **c)** 13 R1   **d)** 10 R4

4. **a)** e.g., yes   **b)** 18

5. e.g., $78 = 60 + 18$ or $78 = 80 - 2$

Mid-Chapter (10) Review
pages 358–359

1. 14 groups

2. **a)** 11   **b)** 24   **c)** 4   **d)** 3

3. 13 boxes, 5 boxes left over

4. **a)** 14 R1   **b)** 27 R1   **c)** 11 R5
   d) 24 R1

5. 16 jackets

6. 5

7. e.g., $60 = 6$ tens and it's easy to divide 6 tens by 6 and it's also easy to divide 18 by 6

8. 17 classes

9. e.g., $60 + 18$ is 78; so breaking down 78 and dividing by 3 gives the same answer as dividing 78 by 3; $78 \div 3 = 26$; $18 \div 3 = 6$, $60 \div 3 = 20$; $20 + 6 = 26$

Lesson 5: pages 360–362
Estimating Quotients

1. **a)** e.g., $62 > 5 \times 10$
   b) e.g., about 12

2. **a)** e.g., about 20 pages
   b) e.g., about 10 pages
   c) e.g., if you have twice as many days, you can read half as many each day. You need to figure out what to multiply by 2 to get 20.

3. **a)** e.g., about 15   **b)** e.g., about 11
   c) e.g., about 12   **d)** e.g., about 16

4. D

5. **a)** e.g., $9 \times 10 = 90$ and 96 is close to 90; or 96 is close to 100 and 9 is close to 10, and $10 \times 10 = 100$
   b) e.g., $8

6. e.g., about 90 pages

7. e.g., any number between 40 and 80

8. e.g., about 10 because the difference between 49 and 68 is almost 20, and the difference between the quotients will be half that amount

9. e.g., about 80 ÷ 2 = 40; between 70 ÷ 2 = 35 and 80 ÷ 2 = 40

Lesson 6: pages 364–366
Dividing by Sharing

1. a) 12 fish b) e.g., there would be a remainder of 3 when you divide 75 by 6

2. a) 14 R4 b) 14 c) 24 R2 d) 12

3. a) 13 b) 31 R2 c) 13 R2 d) 16 R1

4. a) 85 ÷ 5 = ▨ b) 17 c) e.g., yes

5. 15 students

6. a) 27 stamps b) 18 stamps

7. e.g., 78 ÷ 4 = 19 R2

8. 14 cm

9. e.g., You want to have equal groups. To divide 92 by 4, start by putting 2 tens in each group. There's 1 ten left so regroup it as 10 ones. You then have 12 ones in all, and you can put 3 ones in each group. That makes 2 tens and 3 ones in each group.

Lesson 7: pages 368–370
Solving Problems by Guessing and Testing

1. e.g., 25 samosas, 49 samosas

2. e.g., 22

3. e.g., 93 ÷ 3 = 31

4. e.g., a 6-by-9 rectangle

5. 66, 67, 68

6. 8

7. $12

8. 60 stickers

9. 18 units

10. e.g., Trevor had between 10 and 15 apples. He gives 4 apples to each of his friends and has 1 apple left for himself. How many apples did Trevor have to start with? (Answer: 13)

Chapter 10 Review pages 372–374

1. ▨ × 2 = 64

2. 17 bags with 2 stickers left over

3. a) 12 b) 63 ÷ 5 = 12 R3

4. B

5. e.g., Divide each part by 5.

6. 19 stamps

7. A and D

8. e.g., 23 days

9. about 13 weeks

10. a) 12 R4 b) 12 R3 c) 13 R2 d) 29 R2

11. 12 dragons

12. $13

13. e.g., 13, 25

14. e.g., 29

15. 4

Chapter 11
3-D Geometry

Lesson 2: pages 382–384
Recognizing Triangular Prisms

1. e.g., It has 2 opposite faces that are triangles.

2. C

3. **a)** yes **b)** yes **c)** yes

4. e.g., Both are 3-D objects that have all flat faces that are mostly rectangles. One has all rectangular faces and the other has 2 faces that are triangles.

Lesson 3: pages 386–388
Communicating about Prisms

1. e.g., Sort them into 2 groups: objects with at least one square face and those with no square faces.

2. **a)** e.g., My pencil box is a long rectangular prism with 4 faces that are long rectangles and 2 faces that are squares.

3. e.g., A door stopper is a triangular prism.

4. **a)–b)** e.g., 2 groups: 6 faces (A, D, F, G) and 5 faces (B, C, E, H)

5. **a)** e.g., There are 6 faces. Each face is a rectangle, but no faces are squares.

6. e.g., rectangular prisms because most boxes, books, and CD cases are rectangular prisms

7. e.g., number of faces, shapes of faces

Chapter 11 Review pages 391–392

1. **a)** e.g., Both are triangular prisms and have 5 faces, 6 vertices, and 9 edges.
   b) e.g., The faces are different sizes and shapes.

2. **a)–b)** e.g., 2 groups: objects with a triangular face (B, C, E, G) and objects without a triangular face (A, D, F, G)
   c) e.g., a triangular prism with a triangular base that has two edges equal

3. **a)** e.g., a triangular prism made by stacking pattern blocks
   b) e.g., a rectangular prism made with modelling clay

4. **a)–b)** a triangular prism; a cube (rectangular prism)

Cumulative Review: Chapters 8–11
pages 394–395

1. D	4. A	7. B	10. C
2. A	5. A	8. C	11. D
3. B	6. A	9. C	

Glossary

Instructional Words

C

calculate [*calculer*]: Complete a mathematical operation; compute

clarify [*clarifier*]: Make a statement easier to understand; provide an example

classify [*classer* ou *classifier*]: Put things into groups according to a rule and label the groups; organize into categories

compare [*comparer*]: Look at 2 or more objects or numbers and identify how they are the same and how they are different (e.g., Compare the numbers 625 and 526. Compare the size of the students' feet. Compare 2 shapes.)

construct [*construire*]: Make or build a model; draw an accurate geometric shape (e.g., Use a ruler and a protractor to construct an angle.)

create [*inventer* ou *créer*]: Make your own example

D

describe [*décrire*]: Tell, draw, or write about what something is or what something looks like; tell about a process in a step-by-step way

draw [*dessiner*]: 1. Show something in picture form (e.g., Draw a diagram.) [*tirer*]: 2. Pull or select an object (e.g., Draw a card from the deck. Draw a tile from the bag.)

E

estimate [*estimer*]: Use your knowledge to make a sensible decision about an amount; make a reasonable guess (e.g., Estimate how long it takes to cycle from your home to school. Estimate how many leaves are on a tree. What is your estimate of $3210 + 789$?)

evaluate [*évaluer*]: Determine if something makes sense; judge

explain [*expliquer*]: Tell what you did; show your mathematical thinking at every stage; show how you know

explore [*explorer*]: Investigate a problem by questioning, brainstorming, and trying new ideas

extend [*prolonger*]: 1. In patterning, continue the pattern [*généraliser*]: 2. In problem solving, create a new problem that takes the idea of the original problem further

G

guess and test [*essais et erreurs*]: Determine the solution to a problem by guessing possible solutions, testing their accuracy, and then guessing other possible solutions, using the information from previous guesses

J

justify [*justifier*]: Give convincing reasons for a prediction, an estimate, or a solution; tell why you think your answer is correct

L

list [*dresser une liste*]: Record thoughts or things one below the other

M

measure [*mesurer*]: Use a tool to describe an object or determine an amount (e.g., Use a ruler to measure the height or distance around something. Use balance scales to measure mass. Use a measuring cup to measure capacity. Use a stopwatch to measure the time in seconds or minutes.)

model [*représenter* ou *faire un modèle*]: Show using objects and/or pictures (e.g., Model a number using base ten blocks.)

P

predict [*prédire*]: Use what you know to work out what is going to happen (e.g., Predict the next number in the pattern 1, 2, 4, 8, ….)

R

record [*noter*]: Set down work in writing or pictures

relate [*établir un lien* ou *associer*]: Describe a connection between objects, drawings, ideas, or numbers

represent [*représenter*]: Show information or an idea in a different way that makes it easier to understand (e.g., Draw a graph. Make a model. Create a rhyme.)

S

show (your work) [*montrer son travail* ou *présenter sa démarche*]: Record all calculations, drawings, numbers, words, or symbols that make up the solution

sketch [*esquisser*]: Make a rough drawing (e.g., Sketch a picture of the field with dimensions.)

solve [*résoudre*]: Develop and carry out a process for finding a solution to a problem

sort [*trier* ou *classer*]: Separate a set of objects, drawings, ideas, or numbers according to an attribute (e.g., Sort 2-D shapes by the number of sides.)

V

visualize [*imaginer*]: Form a picture in your mind of what something is like; imagine

Mathematical Words

2-D shape [*figure* (f) *à 2 dimensions*]: A shape that has the **dimensions** of length and width

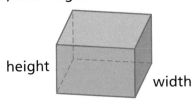

length

width

3-D object [*figure* (f) *à 3 dimensions*]: A shape that has the **dimensions** of length, width, and height

height

width

length

24-hour time [*affichage* (m) *sur vingt-quatre heures*]: A method of telling the time in which the day runs from midnight to midnight and is divided into 24 hours, numbered from 0 to 23 (e.g., Eight o'clock in the morning is written 08:00 and eight o'clock at night is written 20:00.) On the 24-hour clock, 13:00 means 1:00 in the afternoon.

A

addend [*terme* (m)]: A number that is added to another number

a.m. [*a. m.* ou *avant-midi*]: From midnight to before noon

analog clock [*horloge* (f) *analogique*]: A clock that measures time using rotating hands

area [*aire* (f)]: The number of identical objects (**area units**) needed to cover a surface completely

array [*arrangement* (m)]: A rectangular arrangement of items or pictures in rows and columns (e.g., An array can show that 4 × 3 and 3 × 4 have the same **product**.)

 This array shows 4 rows of 3 or 4 × 3. It also shows 3 columns of 4 or 3 × 4. In both cases, the product is 12.

attribute [*propriété* (f) ou *attribut* (m)]: A characteristic or quality, usually of a pattern or geometric shape (e.g., Some common attributes of shapes are size, colour, texture, and number of **edges**.)

axis (plural is axes) [*axe* (m)]: A horizontal or vertical line in a graph, labelled with words or numbers to show what the bars or pictures in the graph mean

B

bar graph [*diagramme* (m) *à bandes*]: A way to show **data** that uses **horizontal** or **vertical** bars

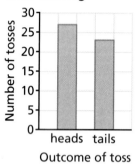

Coin Tossing

base [*base* (f)]: In a 3-D object, the face that it rests on; the 2 **polygon** faces that are the same size and shape in a **prism** are the bases.

 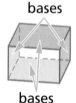

bases   bases

bases   bases

benchmark [*repère* (m)]: A familiar number or measurement to use for comparing other numbers or measurements (e.g., If you compare two fractions by thinking about whether they are greater or less than $\frac{1}{2}$, you are using $\frac{1}{2}$ as a benchmark.)

C

Carroll diagram [*diagramme* (m) *de Carroll*]: A chart with rows and columns that shows relationships

centimetre (cm) [*centimètre* (m)]: A unit of measurement for **length**; one hundredth of a metre (e.g., A fingertip is about 1 cm wide.)
1 cm = 10 mm, 100 cm = 1 m

column [*colonne* (f)]: A set of items lined up vertically; a **vertical** set of squares in a grid (See also **row**.)

column

cube [*cube* (m)]: A **3-D object** with 6 square faces

cylinder [*cylindre* (m)]: A **3-D object** with 2 circular **bases** and one curved side

D

data [*donnée* (f)]: Information gathered in a survey, in an experiment, or by observing (e.g., Data can be in words like a list of students' names, in numbers like quiz marks, or in pictures like drawings of favourite pets.)

decimal [*nombre* (m) *décimal*]: A way to describe fractions using place value; a decimal point separates the ones place from the tenths place (e.g., 0.6 is a decimal for $\frac{6}{10}$ or 6 tenths.)

434

denominator [*dénominateur* (m)]: The number below the bar in a fraction; it tells the number of equal parts in one whole

$$\frac{1}{4}$$

The whole is divided into 4 equal parts.

diagonal [*diagonale* (f)]: In a **2-D shape**, a diagonal can join any 2 **vertices** that are not next to each other

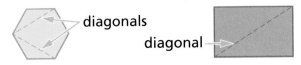

diagonals

diagonal

difference [*différence* (f)]: The result when you subtract; the amount by which one number is greater than or less than another number

$$\begin{array}{r} 93 \\ -\ 45 \\ \hline 48 \end{array}$$ ← difference

digit [*chiffre* (m)]: A symbol used to write a numeral; the digits are 0, 1, 2, 3, 4, 5, 6, 7, 8, and 9 (e.g., For 5608, the digits are 5, 6, 0, and 8. For 7.9, the digits are 7 and 9.)

dimension [*dimension* (f)]: Measurements such as length, width, and height

dividend [*dividende (m)*]: The starting number in a division operation

$$9 \div 3 = 3$$

dividend

divisor [*diviseur* (m)]: The number you divide by in a division operation

$$9 \div 3 = 3$$

divisor

double [*doubler un nombre*]: Add a number to itself (e.g., Double 28 is 28 + 28 = 56.)

edge [*arête* (f)]: The line where 2 **faces** meet on a 3-D object

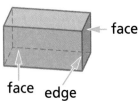

face

face   edge

equation [*égalité* (f) ou *équation* (f) (si inconnue)]: A mathematical sentence in which the value of the left side is the same as the value of the right side (e.g., 1 + 3 = 4)

estimate [*estimation* (f)]: A reasoned guess about a measurement or answer

even number [*nombre* (m) *pair*]: A number that can be divided evenly by 2 (e.g., 12 is even because 12 ÷ 2 = 6.)

expanded form [*forme* (f) *décomposée, sous la*]: A way to write numbers that shows the value of each digit

face [*face* (f)]: A **2-D shape** that forms a flat surface of a **3-D object**

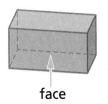

face

fact [*opération (f) mathématique*]: An addition, subtraction, multiplication, or division sentence in which at least 2 of the numbers are 1-digit numbers (e.g., A multiplication fact is 3 × 5 = 15.)

factor [*facteur (m)*]: Any one of the numbers you multiply

fraction [*fraction (f)*]: Numbers used to name part of a whole or part of a set (See also **numerator** and **denominator**.)

Ⓖ

gram (g) [*gramme (m)*]: A unit of measurement for **mass** (e.g., 1 mL of water has a mass of 1 g.) 1000 g = 1 kg

graph [*diagramme (m)*]: A way of showing information so it is more easily understood. A graph can be concrete (e.g., boys in 1 line and girls in another), pictorial (e.g., a picture or symbol to show the number of boys in one 1 row and girls in another), or abstract (e.g., 2 bars on a bar graph to show how many students are boys and how many are girls).

greater than (>) [*plus grand que*]: A sign used when comparing 2 numbers (e.g., 10 is greater than 5, or 10 > 5.)

grouping [*groupement (m)*]: A type of division problem in which a number of items (the **dividend**) is separated into groups of equal size (the **divisor**). The number of groups is the result, or **quotient** (e.g., If 8 items are put into groups of 4, then there are 2 groups; 8 ÷ 4 = 2.)

Ⓗ

halve [*diviser un nombre par 2*]: To divide a number by 2

hexagon [*hexagone (m)*]: A **polygon** with 6 straight sides and 6 angles

horizontal [*horizontal*]: Level with the floor or the bottom of the page (↔)

hour hand [*petite aiguille (f)*]: The short hand on a clock or watch that indicates hours

Ⓚ

kilogram (kg) [*kilogramme (m) (kg)*]: A unit of measurement for **mass** (e.g., A math textbook has a mass of about 1 kg.) 1 kg = 1000 g

Ⓛ

length [*longueur (f)*]: The distance between 2 points, often measured in **centimetres (cm)** or **metres (m)**

less than (<) [*plus petit que*]: A sign used when comparing 2 numbers (e.g., 5 is less than 10, or 5 < 10.)

line of symmetry [*axe (m) de symétrie*]: A line that divides a 2-D shape in half so that if you fold the shape on this line, the halves will match

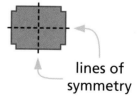

lines of symmetry

Ⓜ

mass [*masse (f)*]: The amount of matter in an object, often measured in **grams (g)** or **kilograms (kg)**

436

metre (m) [*mètre* (m)]: A unit of measurement for **length** (e.g., 1 m is about the distance from a doorknob to the floor.) 1000 mm = 1 m, 100 cm = 1 cm, 1000 m = 1 km

minute hand [*grande aiguille* (f)]: The long hand on a clock or watch that indicates minutes

multiplication [*multiplication* (f)]: A way of adding a number of groups of the same size quickly (e.g., If there are 4 groups of 3, then there are 3 + 3 + 3 + 3 or 4 × 3 = 12.)

net [*développement* (m)]: A 2-D pattern that can be folded into a 3-D object

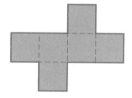

number line [*droite* (f) *numérique*]: A diagram that shows ordered numbers or points on a line

number sentence [*égalité* (f) ou *expression* (f) *mathématique*]: A mathematical statement that shows how 2 quantities are related (e.g., 3 × 8 = 24)

numeral [*nombre* (m)]: The written symbol for a number (e.g., 148, $\frac{3}{4}$, 2.8)

numerator [*numérateur* (m)]: The number above the bar in a fraction; it tells the number of equal parts the fraction represents

$\frac{1}{4}$ ←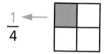

The fraction tells about 1 of the equal parts.

numeric date [*date* (f) *en chiffres*]: A way of writing dates with numbers and no words; one way is year-month-day (e.g., 2002-08-15)

octagon [*octogone* (m)]: A **polygon** with 8 straight sides and 8 angles

odd number [*nombre* (m) *impair*]: A number that has a remainder of 1 when divided by 2 (e.g., 15 is odd because 15 ÷ 2 = 7 R1.)

organized list [*liste* (f) *ordonnée*]: A way of putting information in order to find all possibilities

pattern [*régularité* (f) ou *suite* (f)]: Something that follows a rule while repeating or changing

pattern rule [*règle* (f) *de la suite*]: A description of how a pattern starts and how it continues (e.g., For the pattern 24, 27, 30, 33, ..., here is the pattern rule: Start with 24 and add 3 each time.)

pentagon [*pentagone* (m)]: A **polygon** with 5 straight sides and 5 angles

perimeter [*périmètre* (m)]: The total length of the sides of a shape

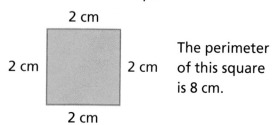

2 cm

2 cm   2 cm

2 cm

The perimeter of this square is 8 cm.

pictograph [*pictogramme* (m)]: A way of showing information that uses pictures or symbols to represent quantities

Rattlesnakes Studied

Female 🐍🐍🐍🐍🐍

Male   🐍🐍🐍🐍🐍🐍🐍🐍🐍🐍🐍🐍🐍

Each 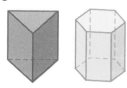 means 5 snakes.

place value [*valeur* (f) *de position*]: The value that is given to a digit based on its position in a number (e.g., The 2 in the number 327 represents 2 tens, while in the number 5.2, it represents 2 tenths.)

p.m. [*p.m.* ou *après-midi*]: From noon to before midnight

polygon [*polygone* (m)]: A closed **2-D shape** with sides made from straight lines

prism [*prisme* (m)]: A 3-D object with opposite faces that are **polygons** of the same size and shape, and other faces that are rectangles

product [*produit* (m)]: The result when you multiply

2 × 6 = 12

factors   product

pyramid [*pyramide* (f)]: A 3-D object with a **polygon** for a base; the other **faces** are triangles that meet a single **vertex** (e.g., a rectangle-based pyramid)

Ⓠ

quadrilateral [*quadrilatère* (m)]: A **polygon** with 4 straight sides and 4 angles

quotient [*quotient* (m)]: The whole number result you get when you divide

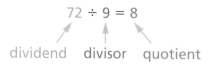

72 ÷ 9 = 8

dividend   divisor   quotient

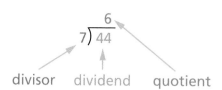

6
7⟌44

divisor   dividend   quotient

ℝ

rectangle [*rectangle* (m)]: A **quadrilateral** with 4 square corners

438

rectangular prism [*prisme* (m) *rectangulaire*]:
A **prism** that has all rectangular faces

regroup [*regrouper*]: Trade 10 smaller
units for 1 larger unit, or 1 larger unit
for 10 smaller units

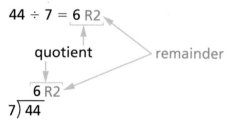

regular polygon [*polygone* (m) *régulier*]:
A **polygon** with all sides the same and all
angles the same (e.g., a square)

remainder [*reste* (m)]: The amount left over
after a number is divided into a whole
number of equal parts

$$44 \div 7 = 6 \text{ R2}$$

quotient   remainder

$$6 \text{ R2}$$
$$7 \overline{\smash{)}44}$$

row [*rangée* (f)]: A set of items lined up
horizontally; a horizontal set of squares
in a grid (See also **column**.)

row

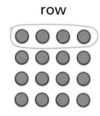

scale [*échelle* (f)]: The number represented
by each unit or shape in a graph

Each  means 5 snakes.

sharing [*partage* (m)]: A type of division
problem in which a number of items (the
dividend) is shared equally among a
number of groups (the **divisor**). The
number of items in each group is the
result, or **quotient** (e.g., If 8 items are
shared among 2 groups, then there are
4 items in each group; 8 ÷ 2 = 4.)

skip count [*compter par bonds*]: To count
without using every number, but
according to a pattern rule
(e.g., counting by 5s: 0, 5, 10, 15)

square [*carré* (m)]: A **quadrilateral** with
4 equal sides and 4 square corners

square centimetre (cm²) [*centimètre* (m) *carré*]:
A unit of measurement for **area**

1 cm

1 cm

An area of 1 square centimetre (1 cm²) is
the amount of space covered by a square
with sides 1 cm long.

square metre (m²) [*mètre* (m) *carré*]: A unit of measurement for **area**

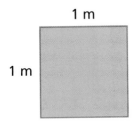

1 m

1 m

An area of 1 square metre (1 m²) is the amount of space covered by a square with sides 1 m long.

square unit [*unité* (f) *carrée*]: A square-shaped unit for measuring **area** (e.g., This design covers 14 square units.)

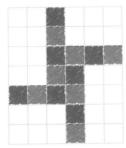

standard form [*chiffres, en*]: The usual way we write numbers

sum [*somme* (f)]: The result when you add

14 + 37 = 51
　　　　↑
　　　sum

survey [*sondage* (m)]: 1. A set of questions planned to obtain information directly from people
[*faire un sondage*]: 2. To ask a group of people a set of questions

symmetrical [*symétrique*]: A way of describing a 2-D shape with at least one **line of symmetry**.

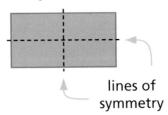

lines of symmetry

table [*tableau* (m)]: A way to present information in columns and rows

tally chart [*tableau* (m) *de fréquence*]: A way to record tallies (e.g., If you are surveying students about whether they watched the hockey game last night, you could use a tally chart like this one.)

Did You Watch the Hockey Game Last Night?

yes	ℍℍ ℍℍ ℍℍ
no	ℍℍ ℍℍ ℍℍ ℍℍ ℍℍ

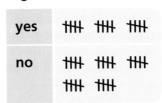

title [*titre* (m)]: A short phrase of text at the top of a **bar graph**, **organized list**, **pictograph**, or **tally chart** that describes what is being recorded or displayed

triangle [*triangle* (m)]: A **polygon** with 3 straight sides and 3 angles

triangular prism [*prisme* (m) *triangulaire*]: A **prism** with triangles as **bases**

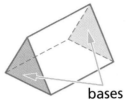

bases

V

Venn diagram [*diagramme* (m) *de Venn*]: A diagram that uses shapes such as circles to show relationships

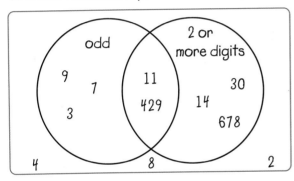

vertex (plural is vertices) [*sommet* (m)]: The point at the corner of a shape or object where sides or edges meet (e.g., A cube has 8 vertices.)

vertex

vertical [*vertical*]: Going straight up from the floor, or up and down ($\updownarrow$)

vertical axis [*axe* (m) *vertical*]: The axis going up and down that is level with the side edge of a page

W

whole numbers [*nombres* (m) *entiers*]: The counting numbers that begin at 0 and continue forever: 0, 1, 2, 3, ….

Index

Credits